The Water Book

Judith Thornton (DPhil)

illustrated by Paul Bullen

Centre for
Alternative
Technology
Publications

Judith Thornton DPhil

© August 2005, 2nd edition 2007

Centre for Alternative Technology, Machynlleth, Powys, SY20 9AZ, UK

• **Tel.** 01654 705980 • **Fax.** 01654 702782

• **email** info@cat.org.uk •**website.** www.cat.org.uk

Illustrations: Paul Bullen

ISBN: 978-1-902175-49-2

1 2 3 4 5 6 7 8 9 10

Printed on 100% recycled Cyclus Print paper in Great Britain by Antony Rowe Limited, 01249 659705.

Published by CAT Publications, CAT Charity Ltd. Registered charity no. 265239.

About the author

Judith Thornton (DPhil) is a water and sewage consultant with Water Works – Wales Water and Sewage Services (www.water-works.org.uk), based in mid-Wales. She has worked with the Centre for Alternative Technology since 1999, until recently as Head of Biology, and currently as tutor on the MSc Architecture: Advanced Environmental and Energy Studies. Judith has installed a variety of systems, including reed beds, compost toilets and private water supplies. Here she brings her vast experience from the field and as advisor on small scale water systems to *The Water Book* for CAT Publications.

Acknowledgements

Judith gratefully acknowledges the contributions made by colleagues at CAT, particularly Louise Halestrap, Dave Zammit and Peter Harper, and the team in the publications department. Thanks to my housemates who fed me coffee and doughnuts to keep me going. Thanks are also due to those who reviewed chapters, including Chris Hartley, Chris Laughton, Peter Harper and Cath Hassell.

The following individuals and organisations kindly provided material for the illustrations:
Environment Agency
Building Research Establishment
Opella
Green Building Store
Hansa
Intermediate Technology Development Group
Allspeeds Ltd
Amos Pumps
Green Shop
Fairey Ceramics
Avonsoft
Meteorological Office
Gramm Environmental
Access Irrigation
DroughtBuster
Water Two

Contents

Introduction

Are you thinking about installing your own water supply? Or do you already have one and want to understand more about it? If your dream is to build a house in an isolated area, the availability of water may well determine whether the project is feasible or not.

At the Centre for Alternative Technology (CAT), we look at all resources (including water) from the viewpoint of sustainability. This book therefore looks at the environmental, social and economic impacts of supplying and treating water from various sources. Whilst it is technically possible to treat almost any water to almost any desired standard, it does not necessarily follow that this is an environmentally friendly or appropriate thing to do, particularly if a water supply infrastructure already exists. Consequently, this book does not advocate self-sufficiency for its own sake. 'Closed loop' systems, in which dirty water that has already been used, is recycled, are considered in light of their energy and resource requirements. Regardless of whether we are using a private water supply or the mains, all of us should concern ourselves with using less water. Increases in environmental awareness and the increasing frequency of hosepipe bans and drought orders have also stimulated interest in how to supplement mains water with water from other sources such as rain.

If, for whatever reason, you are going to rely on your own water supply, you need to know where to look for water. This introduction therefore considers the hydrological cycle, which describes where water is on the planet and how it moves from place to place. It also describes some of the basic properties of water that provide a context to the issues surrounding moving and cleaning it.

You may have to go to considerable trouble to find a water source and to clean it; it therefore makes sense not to use more of it than you need. **Chapter One** will help you calculate your water demand so that you can consider potential supplies in the light of your needs. The water efficiency measures described also provide a good starting point for those of you who have a mains water supply, and are interested in the environmental benefits of using less water.

Armed with a basic knowledge of the hydrological cycle and your water demand, you can examine in detail how to find water in **Chapter Two**. There will often be more than one source available to you, and you will need to consider the advantages and disadvantages of each. An overview of this decision-making process is shown in the diagram below, and the rationale behind it will become clear as you read through the chapters. Once you have found your source, you need to think about how to get it from where you found it, to where you need it and this is dealt with in **Chapter Three**. And, once we have the water where we want it, we must consider how to clean it. The method you use will depend upon what needs to be removed from the water and this is considered in **Chapter Four**.

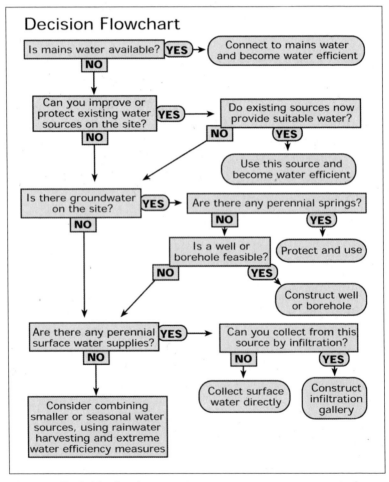

Figure 0.1 The decision flow chart.

Chapter Five deals specifically with rainwater as a source, and **Chapter Six** considers grey water recycling, both of which are additional water sources that you may be looking at if water is in short supply. **Chapter Seven** considers the use of water in the garden – this can make up a considerable proportion of household water use in the summer. These three chapters are designed to stand alone and will be of interest to those who already have a mains water supply but are looking to supplement it.

Case studies are provided throughout the book to illustrate specific points, and to provide examples of how the principles can be translated into practice. Suggestions for further reading related to the content (and details for titles referred to) are given at the end of each chapter.

The book is predominantly aimed at the situation in the UK; solutions appropriate in other countries may not be applicable here for various reasons including water resource distribution and availability, politics, economics and social factors. This is true particularly of the more drastic water supply and recycling options that are the norm in countries with scarce water supplies, but are not necessary in the UK. This book does not consider in any detail either the quality of tap water or the global politics surrounding water supply. Nor is this book about the processes of cleaning water after it has become sewage, for which the reader is referred to *Sewage Solutions* (see Further Reading).

Water: the basics

The hydrological cycle – the way water flows

Water is continually moving between the land, the sea and the air. The cycle in which it moves is called the hydrological cycle, and is illustrated in Figure 0.2 on page 4. The energy driving the cycle comes from the sun. Water in the sea and on the land surface, evaporates and forms clouds. Movement of water from land to air is supplemented by *transpiration* – the process by which plants lose moisture to the atmosphere – so the hydrological cycle is closely linked to ecosystems. In the atmosphere, water forms clouds and these deposit their water back on the land as rain, snow or hail. This water collects in streams and rivers, and flows back into the sea. Some water soaks into the ground, and percolates through soil and rocks: this is known as groundwater, and it may return to the surface in lakes, streams or rivers, or it may return to the sea directly. This is the basic hydrological cycle, but you should bear in mind a few complications. Firstly, there are a number of points at which water may miss out parts of the cycle, for example when it rains over the

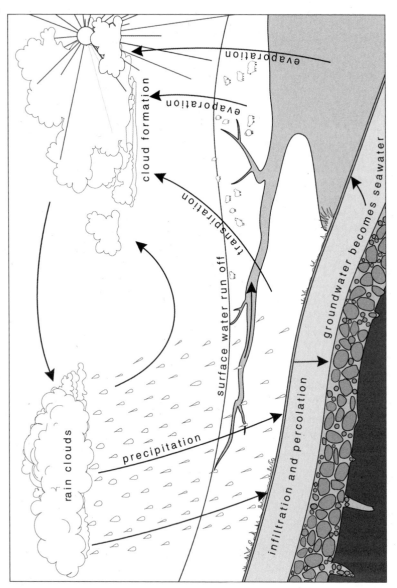

Figure 0.2 The Hydrological Cycle.

Location	Volume Tm^3	% total water
Sea	1320000	97.25
Polar ice caps, glaciers, snow	29200	2.1
Saline lakes and inland seas	105	0.008
Atmosphere	13	0.001
Groundwater	8250	
Soil moisture	65	0.62
Rivers	1.25	
Fresh water lakes and reservoirs	125	

Table 0.1 Distribution of water on the planet.
A note on units: One $Tm^3 = 10^{12}\ m^3$, and there are 1000 litres in a m^3 of water.

sea (this actually accounts for about 90% of total global precipitation). Secondly, the rate of water movement through the cycle varies. These variations can be both seasonal and geographical. In tropical countries, water may evaporate into the sky within minutes of falling as rain, conversely water can be trapped in ice caps for hundreds of thousands of years. Some groundwater is referred to as 'fossil water' since it has been underground for hundreds of thousands of years and would take similar times to be replenished if it were abstracted.

How much water is there and where is it?

The amount of water on the planet is constant, but is moving continually through the hydrological cycle. At any one time, the vast majority of water (97%) is in the sea. The table above indicates the total amount of water on the planet, and the amounts of water in each phase of the hydrological cycle at any time.

Whilst sea water accounts for the largest percentage of water on the planet, turning it into fresh water is an energy intensive process that is best suited to large scale (i.e. thousands of households) applications in otherwise water-scarce countries (such as the middle East). It is certainly not an ecologically sound method to attempt on a domestic scale in the UK; we are more interested in groundwater, soil moisture, rivers and lakes. As you can see from the table, just 0.62% of the planet's water is available to us from these sources.

Where did water come from originally?

Water has been present in the solar system since the Big Bang, having originated during the thermonuclear fusion events that formed the original elements and compounds in the universe. Its presence on earth is as a result of water vapour being given off by the earth's mantle during volcanic eruptions and movement of the mantle several billion years ago. Volcanic eruptions continue to release minor quantities of water into the hydrological cycle today, although this probably amounts to only about a km^3 of water on the planet each year. Comets also contribute small amounts of 'new' water to the earth's hydrological cycle; they are made up largely of ice that vaporises and becomes atmospheric water as they near the earth.

The chemistry of water

Water consists of two hydrogen atoms bonded to one oxygen atom: H_2O. It has some unusual properties owing to the existence of the hydrogen bond, which describes the tendency of one side of a water molecule to be slightly positively charged and the other side slightly negatively charged. This causes them to stick together, so water is liquid at ambient temperatures in the UK (unlike other compounds of similar molecular weight, which tend to be gases). Water is also an excellent solvent, meaning that other molecules such as gases, salts and minerals are easily carried by it. Unfortunately, for us, this means water gets dirty very easily. This 'dirt' will include single-celled creatures that make use of the nutrients in the water. In addition, the specific heat capacity of water is very high; put simply, this means it takes a lot of energy to heat up water by small amounts. This is also advantageous to the creatures that live in water, as it makes it an easier and more constant environment to live in than air.

How much do we need?

Our bodies consist of about 60% water. We are continually losing this from our skin, through exhaling, and as urine and faeces. Public health guidelines suggest drinking around 2-3 litres per day, but the amount you actually need varies with age, food intake, temperature, exercise and type of beverage. Development agencies regard 5 litres per person per day as the minimum for cooking and drinking purposes, rising to 15 litres if water is also used for bathing and clothes washing. If water-borne sanitation is required, this figure increases to 50 litres. As you will see in the next chapter, in the developed world most of us use far more than this.

The taste of water: what's in it and what shouldn't be

Totally pure water with nothing dissolved in it doesn't taste nice and isn't good for you: for example the calcium and magnesium carbonates that cause water hardness are useful minerals for bone growth. Cleaning water isn't always about taking 'bad' things out of it; the process can consist of putting things in. Lots of things in water alter its flavour, often for the better, but in quite abstract ways. Water may taste 'flat' or 'energetic' according to the amount and type of dissolved gases in it. Sometimes water tastes almost sweet, sometimes sharp. The public water supply is regulated by the Drinking Water Inspectorate (DWI) and limits for over 50 parameters (including a range of chemicals, metals and bacteria) are checked. We are exceptionally lucky in the UK that drinking water is so safe; in 2005 almost 3 million tests were carried out on the public water supply and 99.96% were passed. But even if it is safe, you may not like the taste of it. If you don't find the taste of tap water to your liking, the section in Chapter Four on additional treatments for public water supplies might help.

More and more people are drinking bottled water, despite the fact that it is often several thousand times more expensive than tap water. You may prefer the taste, but it is no better or worse for your health than what comes out of your kitchen tap. Bottled water is clearly worse in terms of its environmental cost; it may be pumped out of the ground at an unsustainable rate, put into a bottle whose production consumes more water than the bottle itself contains, and then transported half way across the world.

Further Reading

- A draft report of the published scientific literature on the impacts of water on health. Commissioned by Water UK, July 2002. Detailed study of the importance of water in the body. Downloadable from the website: www.water.org.uk
- Drinking Water Inspectorate – UK government organisation responsible for the quality of drinking water. Website includes factsheets on water quality: www.dwi.gov.uk
- *Sewage Solutions: Answering the call of nature.* Grant, Moodie & Weedon, CAT Publications, 2002. Available from CAT Mail Order

Chapter 1. Water Use

How much water you use – why use less? – how to use less – bottom line water consumption.

Water efficiency as a first resort

As you saw in the Introduction, there is a limited water resource available to us and we should bear in mind the hierarchical approach to resource management; in order of importance: reduce, reuse, recycle. This is based on the fact that it is generally easier to use less of any resource than to find more of it.

If you have a mains water supply, you should read this chapter and implement water efficiency measures before considering supplementing your water from other sources. In fact, if what you are interested in is your ecological footprint, you should bear in mind that water and sewage probably forms a very small proportion of it (generally less than 2%). Hence you would do well to examine the environmental impacts of other areas of your life, like transport, food and energy consumption before becoming overly concerned about your water use.

Why should I use less?

Having had clean, cheap and plentiful water on tap for several generations, many people have little idea about why they should use less water. There are a number of reasons to consider:

- decreasing the demand for potable (i.e. drinking quality) water
- reducing the impacts of purifying and transporting water
- reducing demand for groundwater and the threat to rivers from over-abstraction
- the effect of climate change on water in the UK
- improving sewage treatment
- reducing water and sewerage bills

Decreasing demand for potable water

45% of total water abstracted in England and Wales is for domestic water purposes (the rest being used by industry and agriculture), and this is the fastest growing proportion of water demand. Many areas (such as the UK South East) are facing increases in population, but are already water-stressed. It stands to reason that the cheapest method for meeting the water needs of an increasing population is for all of us to use a bit less.

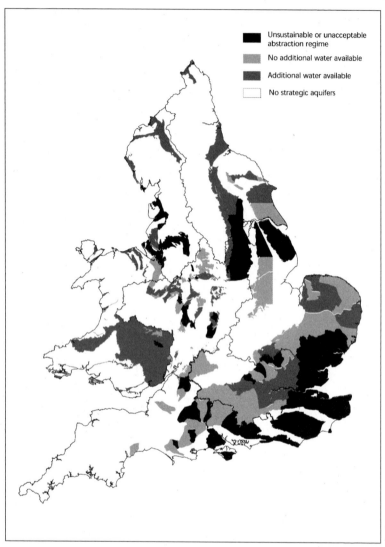

Figure 1.1 *The map shows the current state of groundwater resources in England and Wales. The sustainability of abstractions is monitored by the Environment Agency. In large parts of the country the geology means that there is no groundwater, and in many others there is little more available that isn't already being used (courtesy of the Environment Agency).*

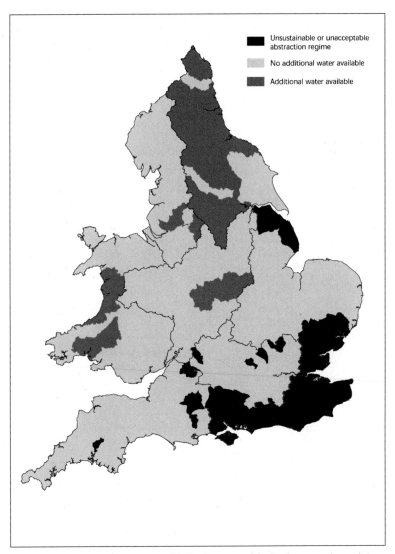

Legend:
- Unsustainable or unacceptable abstraction regime
- No additional water available
- Additional water available

Figure 1.2 Summer surface water availability (courtesy of the Environment Agency).

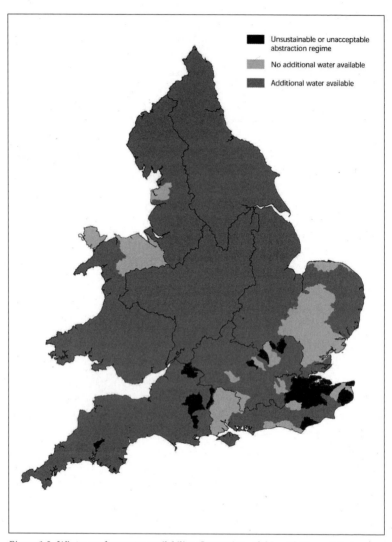

Figure 1.3 *Winter surface water availability. Comparison of this with Figure 1.2 shows how important interseasonal water storage is in many areas: most of England and Wales have abundant water resources during winter, but this must be stored to meet demand in the summer. This is difficult given the public's unwillingness to see new reservoirs being built (courtesy of the Environment Agency).*

Reducing the impact of purifying and transporting water

The water that comes out of your tap will have been through a number of processes to make it clean and safe. This can include screening and filtering to remove solids, coagulation and flocculation to remove colour, and disinfection to inactivate pathogens. Clearly, all of these processes have an environmental impact and reducing water use will help minimise these impacts. This is particularly important in the summer, when there is less water available and water companies are often forced to use much dirtier water that requires more intensive cleaning. The direct energy cost of purifying water, delivering it to your house and cleaning it after you have made it dirty is equivalent to about 80kilograms of CO_2 emissions per household per year.

Reducing the impacts of water abstraction

Aquatic habitats are very sensitive to the amount of water that is in them, so flow and water level variability must be protected. In many of the areas of England and Wales, we are removing water from ground- and surface water at unsustainable rates already, and in others there is no additional water available (Figures 1.1, 1.2, 1.3). Whilst there is still a plentiful surface water resource in many areas during winter, massive capacity is required to store this water for use in the summer, and this stored water then needs to be moved over long distances to supply cities.

The effect of climate change on water availability in the UK

Predictions of the severity of climate change are full of uncertainty and controversy. However, regardless of the average temperature increase that will occur, scientists are in little doubt that weather patterns in the UK will become more extreme. In terms of water, this will mean higher winter rainfall, increased flooding and lower summer rainfall. This will increase the need for inter-seasonal water storage in large reservoirs and will necessitate a change in our approach to stormwater drainage.

Improving sewage treatment

The water that you use in the house ends up as sewage, so generating less sewage will decrease your environmental impact. You probably use around 50 litres of water each day just to flush the toilet; water is a convenient transport medium that can carry solids with it along pipes to where they can be treated. Sewage treatment is then required to remove the solids from the water again. Consequently, adding the bare minimum of water to the waste that is necessary to move it along pipes allows the cleaning process to be easier, cheaper and have a lower environmental impact.

Reducing water, energy and sewage bills

In the UK, most domestic properties (77%) are not metered; you pay for water according to the (old) rateable value of the property. Water metering is becoming more common: new domestic properties and the majority of commercial and public buildings are now metered. If you are on a mains water supply, the water company has the right to insist on installing a water meter if you are in a water-scarce area or have large water uses (such as for garden sprinklers or a swimming pool). Studies by water companies have shown that people on water meters use less water than those who aren't, and pay correspondingly lower bills; these savings can easily be in the order of 30% where meters are installed voluntarily, falling to 10% where meters are compulsory. You can ask your water company to install a water meter free of charge. Your sewage bill will also decrease, since, once you are on a meter, the sewage charge will be calculated on your mains water usage.

Being efficient with hot water will save you money on your energy bills as well as your water bill; often the cost of heating water is more than the cost of the water itself. To heat 40 litres of water (typical individual hot water use per day) from 10^{0}C to 60^{0}C requires 2.33 KWh of energy, which costs between £19 and £60 per year, depending on the heating system. This volume of water will cost around £10 annually.

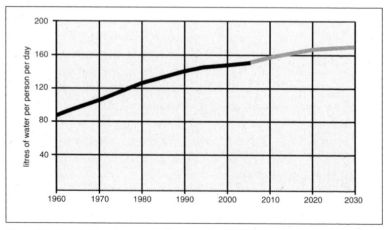

Figure 1.4 Household water consumption in the UK (courtesy of the Environment Agency).

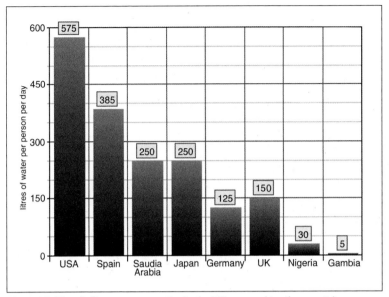

Figure 1.5 Household water consumption in the UK compared to other countries (courtesy of the Environment Agency).

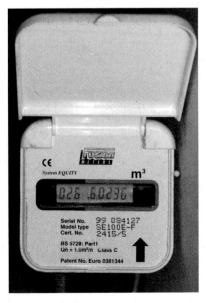

How much water do we use and where does it all go?

Domestic water use per capita in the UK has risen from 100 litres/person/day in 1970 to 150 litres/person/day in 2004/5 (Figure 1.4). This increase is due to a number of factors including smaller households (per captia use is higher in small households than larger ones), cultural shifts towards increased washing and bathing,

Figure 1.6 Domestic water meter. Meters installed by water companies measure the volume of water used in m^3. A cubic metre of water is 1000 litres, so since installation of this water meter, the householders have used 26.6 x 1000 litres of water, i.e. 26,600 litres.

and increasing affluence – resulting in higher numbers of washing machines, dishwashers and larger gardens. The domestic water use of the UK is compared to that of other countries in Figure 1.5. In future, domestic water consumption may decrease as new build houses are given sustainability ratings under the new 'Code for Sustainable Homes'. This awards minimum points for a water consumption of 120 litres/person/day and maximum points for a consumption of less than 80 litres/person/day.

Average water consumption obviously hides a lot of individual differences. If you have a water meter, it is easy to find out how much water you use, either from your bill or by reading the meter (Figure 1.6). If you don't have a water meter, it's a little more difficult. Bizarrely, the water companies often don't know how much is used within households. This is because whilst they know how much they supply into their distribution network, an unknown amount is lost to leakage during distribution. Depending on which technique is used to calculate leakage different

Calculating your water use

The amount of water that an appliance uses varies considerably according to the manufacturer, and the following are estimates of the average UK situation. Simply work out how many times you use each appliance each week, multiply it by the figure for litres/use in the second column and add up the totals.

Appliance	Litres/use	Uses/week	Total water consumption
Toilet	10[1]	2	
Shower	35[3]		
Power shower	100[3]		
Bath	80		
Washing machine	70		
Dishwasher	35		
Drinking/cooking (per day)	5-10		
Washbasin	5		

Table 1.1 Calculating your water use.

[1] Toilet flush volume varies according to the age of the toilet. The most up to date standard is maximum of 6 litres/flush.

[2] Toilet flushes per week can be estimated from (25 x residents) + (15 x daytime occupants).

[3] Showers and power showers have a wide variation in their water use per minute, and clearly the actual water use will depend upon how long you spend in it. It is recommended that you measure how much water your shower uses in a minute if you want to improve the accuracy of your audit. If you are interested enough to plumb in your own flow meters within the home, you can substitute your own numbers for the numbers of litres per use of an appliance in the table above.

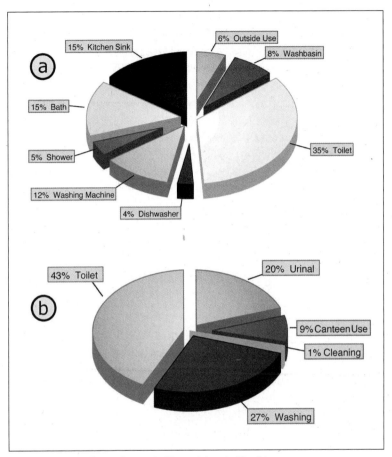

Figure 1.7 Usage of water in a typical house (a), and office (b)
Data courtesy of the Environment Agency and the Building Research Establishment.

results are obtained, but the regulator's (OFWAT) estimate for total leakage during supply in 2003/4 was 3650 mega litres per day (that's 150 litres/property/day!).

You can work out where your water goes by doing a water audit in which you consider all your water-using appliances and how often you use them to give you a total water use, as shown in Table 1.1 on page 16. You may well find that simply by measuring your water use and thinking about the issues a little more, you will naturally start using less water, so carrying out an audit can be an excellent water efficiency measure in itself.

An audit will help to determine where the biggest water savings can be made.

The average domestic situation is shown in Figure 1.7a. If you are interested in minimising water use in your work place, then Figure 1.7b shows the typical water uses in an office. More detailed data on commercial water uses within specific sectors is available via the 'Watermark' project (details are listed in the Further Reading section at the end of this chapter).

How to use less water

There are two ways of using less water. The first is to change your behaviour; the second is to change your water-using appliances.

Behaviour

The following behavioural changes are fairly easy (and free!), and will result in substantial water savings with additional energy savings if the measure is for hot water use.

- Between 5 and 20 litres of water come out of a tap in a minute (depending on water pressure). Don't leave the tap on whilst brushing your teeth, washing etc. You could easily be wasting 15,000 litres of water a year just brushing your teeth!
- Think before you use the hot water tap. Depending on the 'dead legs' in your plumbing system, waiting for it to run warm can waste 10 litres of water. Can you use this cool water for something else?
- When you're using small quantities of water such as for rinsing your hands, use the cold tap instead of the hot. Quite often people use the hot tap automatically, but if it's just a quick rinse then the water probably won't be hot by the time you've finished anyway; all that you will have done is moved a few litres of hot water along a dead leg in the plumbing.
- Use a washing up bowl rather than the kitchen sink. It only needs half as much water to fill.
- Wait for a full load before using the washing machine or dishwasher. The 'half load' or 'economy' settings use at least 75% of the energy and water compared to a full load, so it's not a 'half' load at all.
- Don't cool food under a running tap; kitchen taps can use 30 litres of water a minute.
- Not flushing the toilet after urinating may be acceptable in some circles, but even better is to capture the nutrients in urine by saving it for the garden.
- Measure how much water your shower uses. Switch it off whilst you're using the shampoo and soap.

You may well be able to think of additional water efficiency measures over and above those listed here. There are also lots of behavioural changes you can use to make your garden more water efficient and these are discussed in Chapter Seven.

Modifying water-using appliances

Once you're a master of water-conscious behaviour, you can look at some technical fixes to reduce your water use.

Fixing leaks

Leaking taps or ball valves (such as in a header tank or toilet cistern) can waste enormous amounts of water. A slight drip wastes 30 litres a day and by the time the drips have built up to a trickle, it could easily be 300 litres a day. Both tap washers and ball valve washers are cheap and easily fixed (if you don't know how, refer to a DIY manual; no need to send for a plumber).

Toilets

As you saw from the pie chart in Figure 1.7, a third of our domestic water is flushed down the toilet. Depending on how old your toilet is, it may use up to 12 litres per flush. There are a number of things you can do to your existing toilet to make it more efficient.

• **Put a cistern displacement device in the cistern** (Figure 1.8c): older toilet cisterns are often bigger than they need to be to flush the bowl. You can either put a plastic bottle of water in the cistern, or get cistern displacement devices (often called 'hippos', 'hogs' or 'sava flushes') from your water company. Toilet cisterns marked two gallons or more (i.e. 9 litres) are usually suited to cistern displacement devices. Make sure your toilet still flushes first time with the reduced volume: if you have to flush twice you will be wasting water and should remove the cistern displacement device or try replacing it with a smaller one. If you have a brick in your toilet cistern, it's best to remove it and replace it with a plastic bottle, since over time the brick will crumble and damage the flush mechanism.

• **Modify the flush lever**: retrofit a variable flush lever to your cistern (Figure 1.8b). These interrupt the siphon during the flush cycle, so the water flow stops. Since they operate differently from conventional flush levers, you should make sure that you put written instructions next to the toilet so that people don't waste water by using the wrong setting and flushing more than once.

• **Install a delayed opening float valve**: normally, water starts to refill the cistern whilst it is flushing, and this can waste up to a litre of water every flush. 'Ecofil' inlet valves have a small chamber underneath the valve that delays the cistern refilling until after the flush has finished (Figure 1.8a).

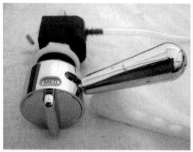

Figure 1.8b Variable flush lever. Flush volume can be changed between 'min','med' or 'max' by twisting the dial on the front. This interrupts the siphon and hence changes how much water is used per flush.

Figure 1.8a Delayed action inlet valve. The chamber underneath the float valve contains water which drains more slowly than the toilet cistern. This prevents the float valve from opening until later in the flush cycle, thereby preventing water wastage (courtesy of Opella).

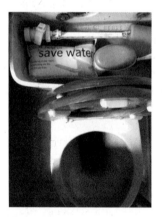

Figure 1.8c A cistern displacement device. These take up space in the toilet cistern so that there is less space available for water, thereby reducing the flush volume.

Figure 1.9 Dual flush toilet.

Other modifications to the flushing mechanism on older toilets are possible (e.g. various types of interruptible flush, in which the flush volume varies according to the time the flush lever is held down). However, these have been found to result in 'double flushing', so increasing water use, unless there is very clear signage about how the flush mechanism works and users are educated as to how the toilet functions.

All of the above measures will help reduce water consumption in your existing toilet, but if you are considering buying a new one, the right choice will reduce flush volume to about 4 litres. A variety of models are available and there are two basic mechanisms available: siphon flush and valve flush. Valve flush toilets can flush with less water than their siphon containing counterparts, and are more likely to have a 'half flush' button, which results in further water saving (Figure 1.9). However, concerns remain about the long-term efficiency of valve flush toilets, as they are more prone to leakage than siphon toilets. If you are the kind of person who will spot a leak and do something about it, get a valve flush toilet, otherwise you should stick to a siphon flush.

Vacuum flush toilets, as seen on planes, boats and trains are also very water efficient, using around 0.5 litres/flush. In most systems the vacuum is continuous, that is to say that the pressure in the pipework is below atmospheric pressure. When the flush button is pressed, a valve opens in the toilet and the contents of the bowl are sucked along the pipe. Clearly it takes energy to generate this vacuum, so this type of toilet is best suited to applications where water use issues are more important than energy issues (e.g. on aeroplanes, where the energy implications of a vacuum flush are minimal compared to the weight implications of the volumes of water required). However, low energy vacuum flush toilets (in which the vacuum is only generated when the flush button is pressed, rather than being continuous) are available, and may be more appropriate to the domestic situation. The most water efficient toilet of them all is a compost toilet, where no flush water is required and you have the added benefit of being able to recover the nutrients and use them in the garden. It is possible to fit a clean, sanitary compost toilet into a domestic bathroom, and the interested reader is referred to *Lifting the Lid* (see Further Reading).

Urinals

Urinals are unusual in the domestic bathroom, but if you have men in the household, and enough space in the bathroom, a waterless urinal can dramatically cut water use. 2 main types of waterless urinal are available: those that include a specifically designed bowl (and therefore tend to be quite expensive), and those that can be retrofitted into a conventional urinal bowl (so are much cheaper).

Unfortunately, many urinals have disposable cartridges, the production and disposal of which may have a higher environmental impact than the water saved.

Showers and baths

Shower flow rates vary a lot, and there is an increasing trend towards power showers. Flow can be anything between around 6 and 25 litres a minute. A bath holds around 80 litres, so you should measure how much water your shower uses before deciding whether or not it is more water efficient than a bath. It is now possible to get low flow shower heads that use around 50% less water than a standard model but get you just as wet. These work by atomising the flow (delivering it in smaller droplets), by aerating the spray (giving the sensation of greater volume) or by having a reduced number of jets. Flow restrictors also exist that can prevent the delivery of more than a set number of litres per minute. If you wish to install a low water-use shower you should consult an ecological builders' merchant that will recommend something compatible with your domestic hot water system; many have minimum requirements for pressure and flow rate. Since our shower habits are often as much to do with sensation and experience (e.g. showering to wake up, invigorating showers) as getting clean, so the choice of shower head and flow rate is often very subjective.

Taps

Between 5 and 20 litres of water come out of a bathroom tap per minute, and up to 30 in the kitchen (the flow rate is dependent on water pressure and the model of tap). The simple behavioural change of turning a tap off will result in substantial water savings, but there are also technological fixes that will reduce water use by taps:

Figure 1.11 Tapmagic devices. These screw into the outlet from your tap. If you turn the tap on half way a spray is produced, but if the tap is turned further a full flow is obtained.

Figure 1.10 Flow restrictor (courtesy of Green Building Store).

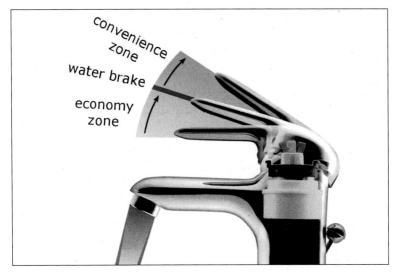

Figure 1.12 Tap with 'water brake'. On pulling the lever upwards, you have to overcome a noticeable resistance (called the 'water brake') to obtain a full flow rather then a low flow (courtesy of Hansa).

• **Flow restrictors** (Figure 1.10): these are plumbed into the pipework leading to the tap and will limit the water to a maximum flow rate, and have added benefits of reducing noise, splashing and 'water hammer' and may help balance the flows within your house if several appliances are being used simultaneously.

• **Shut-off mechanisms**: various mechanisms are available to make taps turn off automatically. These include percussion taps (which you push to turn on, but turn off automatically) and taps with infra-red proximity sensors. Better quality models allow you to adjust the length of time before the tap turns off.

• **Spray fittings**: the continuous flow of water coming out of taps isn't actually a very efficient way of getting wet! Spray fittings can reduce the flow of water through a tap by up to 80%, yet still result in adequate wetting. These are an excellent choice in situations where you don't need to fill the basin/sink (e.g. for hand washing), but adjustable models are also available which allow a normal tap to be converted to one with two settings; both spray and full flow (Figure 1.11).

• **Taps with 'water brakes'**: these are lever taps in which the lever is easy to push to a certain point, and in this range the flow is low (e.g. under 5 litres/minute). To obtain a full flow, you must push the lever further, past a noticeable resistance point (known as the water brake). The tap illustrated in Figure 1.12 (by Hansa) also has an integral adjustable flow restrictor.

Washing machines and dishwashers

The majority of modern washing machines use less than 50 litres per full load. 'Half load' or 'economy' settings are a bit of a con, often only reducing water and energy use by 10-25%, making them a poor substitute for waiting until there is a full load.

Modern dishwashers use 20 litres or less on a full load (of 12 place settings). But how does this compare to washing by hand? A standard washing up bowl filled to about an inch from the top holds approximately 7 litres, and a standard sink holds about 15 litres, so there is no definitive answer. If you tend to rinse things before putting them in the dishwasher, wash the saucepans separately, use the half load function and don't put your best crockery in, then washing up with a dishwasher will not be particularly water efficient. But if you would otherwise wash up under a running tap and this takes you 20 minutes at 12 litres of water per minute, then you'd be better off with a dishwasher. Also, since the major environmental impacts of a dishwasher during its lifetime are associated with detergents, rather than water use, there is probably little to choose between a dishwasher and hand washing.

Garden

Whilst it does not account for a major proportion of our water use on an annual basis (6%), use of water in the garden increases during the summer when demand is naturally high and supplies are most stressed. There are a number of ways in which you can garden in a more water efficient way; these are considered in detail in Chapter Seven.

Water demand for a house on a private supply

Implementing behavioural changes and installing water efficient appliances can reduce your water use from a UK average 150 litres/person/day to 80 litres/person/day fairly easily. Further savings can be made if necessary, and households with compost toilets and who are dedicated to water efficiency can use less than 40 litres/person/day. If you are going to rely on a private water supply, you will need to decide how much, or how little, water you are likely to use, and this will determine what water sources are realistic, how much storage you will need, how to treat the water and how big the pipes will need to be. The sustainability of the water supply itself will also determine how water efficient you need to be. There may be instances where it is acceptable to be fairly profligate, since the water you use will be going back into the same catchment that you abstracted it from and there is plenty there, but in other circumstances (e.g. if you are relying entirely on rainwater harvesting) you may need to be very frugal. You may be happy to make significant lifestyle changes and survive on minimum water use, but if you are planning to expand or sell the property, house more people, or have regular guests, you may not necessarily get the same low level of water use by other people. Convenience can have a major effect on water use; if you have to go to much effort to get water you will probably use less of it (for example if you choose to pump water by hand rather than electrically). How variable is your water demand likely to be? Will the washing machine be in use at the same time as the bath or shower? What are the consequences of your supply running out for a couple of hours or even several days? Do you need to know you'll always have plenty of water or might you be willing to forego the washing machine and garden watering for a few weeks in the summer? Is your water supply likely to be reliable during long dry periods or are you going to have to store water for several months? Sometimes a relatively minor increase in infrastructure cost can substantially increase your storage capacity. You may also decide to have a modular storage and treatment system that will easily allow for increases in future demand. If this is the case you may choose to take a risk and estimate a very low daily water consumption but be willing to increase storage capacity at short notice if you get your estimates wrong. In most cases, the best way of deciding how much water you are going to need is to fit a flow meter on to your water supply and measure your water use over as long a time period as possible.

Virtual water and the water footprint

This book is about the water you use directly. However, no discussion about water efficiency would be complete without some consideration of 'virtual water' and the 'water footprint'. These terms are used to describe the water that is consumed in the production of goods and services. The amount of water used for these purposes vastly exceeds that used directly by you in your home. For example, 140 litres of water are required to grow enough coffee for a single mug full, and 16,000 litres of water are required to produce 1 kilogramme of beef. Whilst these numbers seem worryingly high, in some instances, this use of water can be fairly sustainable e.g. where food is being produced in areas where agricultural water use is efficient and there is plenty of water available. However, there are many exceptions to this; globally, agriculture is by far the biggest water user, and there are widespread examples of where this is at the expense of local populations or where agricultural water use is very inefficient. And it's not just food, water is vital to the production of most goods and services. In the UK, our average individual water footprint (the total amount of freshwater used to provide goods and services for a person) is 1245m^3/year. Compare this to the average domestic water use (around 60m^3/year). Clearly, if you are interested in wider ecological issues rather than simple domestic water efficiency, low flush toilets would not be your hightest priority! Information sources on virtual water and the water footprint are given in Further Reading.

Summary

Water efficiency should be the first measure considered before resorting to alternative supplies. Similarly, behavioural changes are preferable to changes in appliances and can significantly decrease water use from the UK average of around 150 litres/person/day. As the largest consumer of water in the home, the toilet should be the first consideration when trying to upgrade or modify water-using appliances. Water efficient alternatives to existing water-using appliances are available, and may result in other performance benefits.

Further Reading

- Environment Agency website (www.environment-agency.gov.uk) has lots of useful resources including:
  - Water Resources for the Future - future strategies for England and Wales, facts, figures, regional maps of water resources
  - Demand Management Bulletin - news on leakage, education initiatives, water efficiency
  - Saving Water in Buildings fact cards – information on a range of water efficiency products, including suppliers details
- The impact of water efficiency on CO_2 emissions. Client report number 205944, Building Research Establishment – summary of energy use by the UK water industry. Available from www.bre.co.uk
- OFWAT (the regulator that oversees the water companies) website (www.ofwat.gov.uk) collates data on water company performance against a number of indicators
- Which? (The Consumer's Association) regularly reports on the performance of household white goods, including their water efficiency. www.which.net
- UK water companies all have free water efficiency advice leaflets and most provide free cistern displacement devices
- The water efficiency of retrofit dual flush toilets, T. Keating, R. Lawson, Published by the Environment Agency and Southern Water, 2000
- The Millenium Dome 'Watercycle' experiment: to evaluate water efficiency and customer perception at a recycling scheme for 6 million visitors, 2002. S. Hills, R. Birks, B. McKenzie, Water Science and Technology 46:6 pp 233-240

- *Lifting the Lid: An ecological approach to toilet systems*, Harper & Halestrap, CAT Publications, 1999, looks at water efficiency in the context of composting toilet systems. Available from CAT Mail Order
- www.waterfootprint.org – information resource for virtual water and the water footprint
- WATERMARK project: a project aiming to benchmark water use in various sectors. www.watermark.gov.uk

Chapter 2. The Mains and the Alternatives

Sustainability of mains water and drainage – a closer look at the hydrological cycle – assessing a site for a private water supply – quality and quantity of sources – springs, boreholes and surface water – how to get the water out – reinstating old supplies – protecting your water.

Before we consider private water supplies, we need to consider the existing mains water system and the environmental problems associated with it, so that we have a benchmark to compare the alternatives to. A related issue is the sustainability of sewage and rainwater disposal systems; since if you are considering rainwater or grey water recycling for water supply, you will be decreasing your impact on drainage systems. The environmental, economic and social impacts of mains water supplies and drainage systems vary from region to region, so you should consider your local circumstances and your likely private water source if you are interested in which is the most genuinely sustainable option.

Environmental sustainability of water supply

Water supply

Supplying water is an infrastructure intensive business, and much of the environmental cost is associated with the installation of the infrastructure rather than with the cleaning and supply of each litre of water. Installing an alternative water supply, if mains water is available to you, is therefore an extra infrastructure cost and an extra environmental impact. Clearly, there will be cases where the environmental costs of mains water supplies are high, such as where the existing mains network is a considerable distance away from where it is needed, or where the water company is over abstracting from the source. In these situations, a private water supply could be the most environmentally sustainable option. In water-scarce areas you should pay particular regard to water efficiency measures, and installing an alternative supply may be an option if the source you abstract from is less overstretched than that used by the water company. However, since the water sources in a catchment area are linked, this is often not a simple thing to establish.

Environmental sustainability of water and sewage disposal

The sewage that you produce leaves your house and flows along sewers to a treatment works. At the sewage works, nutrients and pathogens are removed before the (now clean) water is discharged back into the hydrological cycle. Unfortunately, in many areas of the UK, rainwater enters the same sewers. Sometimes this is due to infiltration from the ground, where sewer pipes have become cracked, or water enters around the joints. Even worse is the large number of deliberate connections of rainwater and surface water drains into the sewerage system. The consequence is that rainwater gets mixed with domestic sewage and industrial waste water and must be treated as sewage at the sewage treatment works. Additionally, the sewerage network and the sewage works are often not able to deal with the massive increase in flow that occurs during heavy rainfall (the flow within a sewer can increase 10-20 fold during a storm), and this can result in untreated sewage being discharged directly into watercourses in order to avoid flooding the sewage works; this is known as combined sewer overflow. Surface water connections into the sewerage system are no longer permitted, but it will be many years before the mains infrastructure keeps rainwater and sewage separate.

Conventional drainage systems have the additional problem that they collect and channel water from several sources into a single stream, thereby increasing the flow during high rainfall and potentially increasing the risk of flooding. This is completely unlike the natural hydrological cycle, where water would infiltrate into the ground close to where it fell, or flow over a ground surface until it met a watercourse. This is now being mimicked by the installation of sustainable urban drainage systems (SUDS). These systems take the rainwater from buildings and impermeable surfaces, slow down its flow and filter it before discharging it back into the environment. Where the soil is sufficiently permeable, water is discharged to ground via soakaways, thereby replenishing groundwater supplies and mimicking the natural hydrological cycle. In areas where the ground is naturally very wet and water won't soak away, SUDS systems that incorporate tanks or ponds are used to slow the flow, before discharging the water at a slower rate into a watercourse.

Economic sustainability

Given the current charging system for water supply in the UK, installing your own water supply will rarely pay back the initial investment or the running costs if you already have mains water. A number of factors are involved:

- being on a water meter – if you don't pay for your water according to how much you use, installing a private supply clearly doesn't make economic sense

- charging policy of water company:
  - is there a charge for disposal of surface water, and can you expect a rebate on this if you reuse your rainwater?
  - is the sewage charge based on metered water supply? If it is, then using an alternative water source becomes more economic
- demand for non-potable water. Typically this is higher in buildings used by lots of people during the day (when the water demand is largely for toilet flushing), or where water is used for other purposes, e.g. industrial processes. Lower grade water from a private supply may be cheaper than mains water in this instance
- is your project a new build or a retrofit to an existing property? It is often more expensive to retrofit water supplies than to incorporate them in the initial design
- is there a requirement to limit surface water run off during storm events (frequently a requirement of planning permission)? If tanks are required to do this, then it may make sense to reuse this water
- community scale systems can be more economic, as the infrastructure cost can be divided between more households.

Social sustainability

Social aspects of sustainability are notoriously difficult to measure and the following points are far from exhaustive. The installation of private water supplies for demonstration or educational purposes in public buildings may have some merit, since it may result in improved water awareness and encourage wider uptake of water efficiency measures by visitors. However, given the complicated nature of the control units and the risk to health in the event of system failure, the social benefits of private water supply installation are questionable. Indeed, the risk of contracting a disease from a private supply is estimated to be between 22 and 50 times higher than from public mains. If they end up costing the householder more than a conventional mains water supply then they are certainly not appropriate for social housing projects. The necessity for basic maintenance such as cleaning filters and checking tanks can be considered as a cost (because of the time taken) or a benefit (in terms of putting people in touch with their resource use) depending on your viewpoint.

Finding a private water supply – assessing the site

The hydrological cycle discussed in the Introduction indicates the most likely sources of water available to us.

The hydrological cycle and site assessment

Sea water is not an appropriate domestic water supply in the UK. Whilst it is plentiful, purifying it is energy intensive and the resulting water is unpalatable.

Surface water (lakes, streams, rivers, reservoirs and canals) arises either from groundwater (where porous rocks reach the ground surface) or directly from rainfall and run off. If it arises from rainfall then the amount of water can vary considerably according to recent weather and the size of the land area that provides the catchment. The quality of surface water is highly dependent on the weather; run off from the land will contain potentially harmful micro-organisms and lots of soil particles.

Rainwater as a direct source of water for domestic water supplies is considered in Chapter Five.

Fog and dew harvesting are possible, but are best suited to countries with extreme water scarcity and suitable diurnal weather patterns.

Groundwater is surface water after it has soaked into the ground. Water can exist in the ground wherever there is space between rock or soil particles. This can either be when the ground is 'unconsolidated' (like soil, gravel, or sand), or 'consolidated' – rock with lots of small holes or cracks in it (like sandstone, shale or limestone). Groundwater is generally much cleaner than surface water owing to the filtering that occurs as it passes through the ground. Rocks that can hold water and allow it to flow through them are known as aquifers, rocks that do not are aquitards or aquicludes. Materials such as clay are not aquifers, because although they can contain water, it is held tightly to the clay particles so cannot be easily removed.

The movement of groundwater is hard to predict, since water is subject not only to gravity, but also to pressure from surrounding rocks, and to capillary action. The boundaries between rocks of different water holding abilities will rarely be straight lines, so the water table (the uppermost level of ground that is water saturated) may not be horizontal over significant distances. Unless it enters the sea directly through porous rocks on the seabed, groundwater will eventually surface somewhere on land; sometimes these points are springs. Other groundwater will flow directly into a river whose banks are permeable, so the water level in the river is the same as that in the ground. If you drill into an aquifer you may create a water table well, whose level is the same as that of the aquifer it is drilled into. In other cases, the water

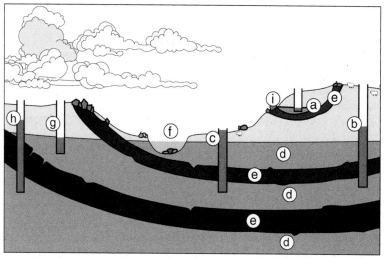

*Fig 2.1 Groundwater. The following features are illustrated. (a) perched aquifer
(b) artesian well (non flowing) (c) artesian well (flowing) (d) aquifer (e) aquitard
(f) stream fed by groundwater (g) water table well (shallow) (h) well/borehole (deep)
(i) spring.*

level in your drilled hole may reach a level higher than the upper boundary of the
aquifer. This is known as an artesian well, and occurs if the aquifer is confined; an
aquiclude above the aquifer prevents the groundwater level rising. The area that an
aquifer refills from is known as the recharge area, and the points/areas from which
it loses water are known as the discharge areas. Figure 2.1 shows some of the critical
features of groundwater, its movement, and ways to access it.

The amount of water available from a groundwater supply depends on a number
of factors. The first is how extensive the aquifer is. If you find a perched aquifer
resulting from small deposits of impermeable rocks forming a basin, it may run
dry regularly. Even if you bore into a large aquifer, you may be removing water
from it faster than it can be naturally recharged. The recharge of the aquifer may be
heavily dependent on seasonal rainfall if it is fairly shallow, but it will also depend
on the hydraulic pressure from water in the surrounding rocks, and how well they
transmit water, i.e. the permeability of the rock. An extreme example of this is clay.
Whilst it contains a large amount of water, it is bound tightly to the clay particles
so the permeability (and therefore rate of recharge) would be very low. At the other
extreme, geology that includes large fissures (such as limestone) will have a very
high permeability. Changes in the amount of water in deep aquifers may take place
over several years as the water may have entered the ground many miles away. Only

if the depth and extent of an aquifer have been completely mapped can you really have any certainty of how much water can be abstracted from it.

Quantity of groundwater will not be your only concern. Its quality will vary according to where the water has been and what it has passed through. Water will be filtered as it passes through rocks, but some (e.g. sand or gravels) provide more filtration than others (e.g. limestone or granite with large fissures). Groundwater will dissolve salts, minerals and gases within the rock. Deep groundwater generally contains very few micro-organisms, since it is difficult for them to survive underground, but will contain more dissolved minerals than shallow groundwater. Man-made chemicals such as nitrate fertilisers and organic pesticides will percolate into groundwater and you will need to know where the aquifer is being recharged from and whether or not that area is at risk of pollution. It is also possible to affect the quality of the aquifer by abstracting water from it; the water that moves in to take its place may have different characteristics, or may be saline.

Looking for water

The best way to assess a site for potential water sources is to consult local experts, and then complement the existing knowledge of a site by assessing it from the ground.

People to ask

Try and find out where your nearest neighbours get their water from, and whether they have had problems with it. If there are wells nearby, find out how deep they are (both resting water level and total depth), how much water they yield and what the water quality is like. Older people will often remember the sites of wells that have since been filled in. Local farmers will know how soil conditions change over the seasons and what the level of the water table is. They may also be able to tell you if any effluent is being discharged into the ground near your site and the location of any land drains. Get in contact with the local Environment Agency Water Resources Officer. You will need to talk to them about abstraction once you have found your source, but they may also tell you what the best local sources are, how much water you'd be allowed to abstract, what's likely to be in it and whether or not there is anybody else abstracting from the source.

If you are going to rely on a groundwater supply, you will probably decide to get a specialist local water engineer to do the work, and they will generally offer to do everything from site assessment to accessing the source and laying the pipe. Have a look at local hydrogeological maps; these are available from the Institute of Hydrology and the Meteorological Office, or from local geology enthusiasts. The amount of detail available will vary, but may mean that you will have a good idea of whether or not groundwater is going to be available before you even set foot on the site. If there isn't a hydrogeological map of the area, it may well be because there aren't any useable aquifers.

Signs on the ground

Have a look around the site armed with as large a scale map as you can get. You will need to work out where water is, where it comes from and where it goes to, i.e. how your site fits into the hydrological cycle. Start by marking out all surface water. Include streams and rivers, pools, and marshy ground. Walk upstream along any surface water to determine whether it comes from a spring or is from rainfall; if it is from a spring it would be better to take water from the spring directly rather than from the surface water, as the surface water has had more opportunity to become contaminated. Have a look at the types of vegetation on the site as these will often indicate water close to, but not reaching, the surface. Plants such as ferns, sedges, rushes and reeds indicate wet ground.

Looking at the site from above can be helpful, either from a nearby hill, or by referring to an aerial map. These may allow you to detect geological features such as fault lines, surface rock features, erosion in places you wouldn't expect it, and tell-tale changes in soil type or the darker lusher green of vegetation that may indicate a spring. They will also allow you to guess how big the catchment area is for any surface water on the site. The catchment area provides some in-built storage, which slows water run off into streams and keeps water flowing when it isn't raining. Bigger catchments will provide more of a buffer than smaller ones, and catchments with steep slopes, exposed rock, shallow soil and little vegetation will have rapid run off and so provide less of a buffer. Are there any land drains in the area? Fields that are artificially drained must channel the water to somewhere else, and may be a useable source of water, although the quality isn't necessarily good and they can run dry in the summer.

Bringing in the experts

If you wish to drill a well or borehole, the chances are you will be getting it built by a professional. Before hiring one, ask about charges. Some don't charge if the hole they drill is dry, others charge per metre of hole drilled. If you know you are getting professionals in to do the job, then you needn't worry about exactly how to find water, as that will be part of the job of the hydraulic engineer.

Dowsing

You may be interested in dowsing for water; that is, searching for underground water by watching the movement of sticks, rods, or pendulums held in the hands.

Those who believe that dowsing works have often had a direct experience of either meeting someone who dowses successfully, or trying it themselves. A believer might say that dowsers are supremely sensitive to some property of water, possibly its conductivity or magnetic effects. The sceptic would say that dowsers look for some of the signs for groundwater discussed above and then subconsciously transmit their hunch to their dowsing rods. The more rigorously conducted studies on dowsing suggest that some individuals have a dowsing success rate far greater than that expected due to chance alone, and very similar to that of any water engineer who has arrived on site with a drilling rig. You may choose to dowse for water, or ask someone to do it for you, but we would caution against relying solely on DIY dowsing before hiring plant machinery to access any water source.

Legal aspects

The aquatic environment in the UK is controlled and monitored by the Environment Agency (Scottish Environmental Protection Agency, SEPA, in Scotland; Environment and Heritage Service, EHS, in Northern Ireland). In general, if you wish to abstract less than 20m^3 (20,000 litres) of water per day from either groundwater or surface water on your own land you will NOT require an abstraction licence. However, the EA have the right to alter this threshold according to local water resource availability, so you will always need to consult them. The EA may require you to carry out specific tests on a groundwater source, to ascertain that the abstraction will be sustainable, before granting any licence. If you wish to abstract from surface water, there will be a minimum amount of water that you must leave in the surface water at all times, and you will have to demonstrate that your abstraction will not result in any damage to the local ecosystems. Licences will usually specify a maximum volume that can be abstracted, and a review date (normally 12 years), after which the licence can potentially be revoked if the sustainability of the local water resource has been altered. The process of balancing the various issues is managed on a catchment by catchment basis using a process known as CAMS (catchment abstraction management strategies). Licence applications can take three months and, if the application is for groundwater, it will often require a groundwater investigation consent and pumping test prior to applying for the licence, so the process can take even longer.

Exclude sources on basis of quality

As soon as you have access to your water supply, you need to make sure that it is of good enough quality to be treatable to the standards you require. Whilst it's technically possible to treat almost any water to drinking standards, that does not mean it is advisable to do so, and in some instances you may be better off finding another source than cleaning one of poor quality. As part of deciding which source you will use, you will need to get the water quality tested professionally, but you can tell quite a lot about the quality of water by looking at it closely.

Is it clear and does it smell? Removing solids is usually the first stage of cleaning water, so the less present initially the better. If you leave the water to settle for a couple of hours you may be able to tell the difference between very small suspended solids and dissolved particles that are colouring the water. Dissolved substances in the water that cause it to be coloured are not always dangerous, but at the very least will influence how you choose to clean your water. They may be due to organic materials leaching out from peat, or decaying vegetation. This may in turn cause algae to grow in the water, which in turn will support a population of other micro-organisms. Colour may also be due to the presence of metals, not necessarily in harmful quantities, but they might not be aesthetically acceptable to you. On occasion, crystal clear water can also indicate that levels of some toxic material are so high that nothing can live in the water, but you won't discover this unless you have the water tested. Avoid water that smells; the dissolved gases that result in smells are not always harmful in themselves, but indicate the presence of other undesirable substances in the water.

Having looked at the water itself, look at where it comes from, as it may give you a good idea of potential contaminants. If you have looked at geological maps, you can predict what types of minerals and dissolved salts could be in groundwater, or you may know that there is a risk of heavy metals from certain types of rocks in the area. Is there a history of mining in the area? Might the ground be contaminated by old, capped-off landfill sites? If it's surface water, what are the surfaces it's being collected from? Are they fields where animals graze, or are they planted and farmed intensively? Is anyone else taking water from the source, and, if so, have they had it tested? Are there septic tanks or land drainage systems in the area or any licenced or unlicenced discharges?

These preliminary indicators of water quality may allow you to discount sources fairly quickly, but once you have narrowed down your choices to a particular source, you should get it tested professionally before you invest in complicated infrastructure to get the water out. How to go about getting your water tested is dealt with in Chapter Four.

Measuring how much is available

Once you have identified a source of suitable quality, you need to determine whether there is enough of it.

Groundwater

The amount of groundwater available from a source is dependent on the size of the aquifer that you have accessed, and the rate at which it is refilled (known as the recharge). Clearly, if you remove water from an aquifer faster than it can be recharged, your source will dry up. Groundwater investigations are best carried out in the summer after a period without rain; if there is water present after a dry spell, you can be reasonably certain that the aquifer will yield water all year. It also reduces the risks of being misled by a perched aquifer (see Figure 2.1), which is highly dependent on seasonal rainfall and which can be plentiful in the winter, but completely dry in the summer.

Wells and boreholes

A basic pump test will determine the likely yield of your aquifer, and is detailed in the box below. These tests should be left to specialists, and the local EA officer may require a specific type of pumping test done before allowing you to abstract water. This is only likely to be necessary for large abstractions, or where significant installation costs and risk are involved.

Pump tests

A pump test is done to establish how fast water flows from the ground into a well or borehole. It requires water to be pumped out in a controlled way, with the resulting effects on water levels being measured. As you remove water, the water level goes down in the immediate area. This effect on water level is known as *draw down*, and the dry(er) area it causes takes the shape of an upside down pyramid, known as a *cone of depression*.

For large abstraction volumes, the impacts of the abstraction on water levels in the surrounding area and on the water environment in general will be measured over a longer period (weeks or months), particularly if there are other abstractions occurring from the same aquifer system. Various types of pump test can be performed depending on what type of data is sought. Samples for chemical analysis may also be taken during the pump test and these can indicate from where the groundwater source is being recharged.

There are lots of factors that affect how fast your groundwater supply will be recharged and where from; how many of these are worth investigating depends upon the importance of your water supply. Will it be a serious inconvenience to you if it runs out periodically? If the consequences and the expense justify it, or if it is required by the Environment Agency, your water specialist may do a range of other investigations on the groundwater.

Springs

The simplest way of measuring the flow of a spring is using the 'jug and stop-watch' method. Simply hold a measuring jug or bucket under the spring and time how long it takes to fill. You may need to dig around the spring outlet a bit to make it easier to funnel all the water into the measuring jug or bucket.

Surface water

It is worth checking with the local EA and the Centre for Ecology and Hydrology (formally the Institute of Hydrology) whether or not they have flow data on local streams, as this is done on many watercourses in the UK for flood monitoring purposes (look in the National River Flow Archive). Even if they don't have it for the stream you are looking at, they will have helpful advice on how much the flow varies according to the season in other local watercourses. You can then measure flow in your stream regularly over a period of time, compare it to other local flow data and rainfall, and then estimate how it will vary over the course of a year. More sophisticated analysis is possible using computer programs containing hydrological mapping data for the whole of the UK. Detailed consideration of measuring stream flows is available in *Going with the Flow* (see Further Reading).

Getting the water out

Once you have ascertained that there's enough water and it's of a suitable quality, you need to determine how best to get it from your source into a pipe and from there to your house. It's important to keep the water as clean as possible during this process, not just so that you don't have to clean your water more than absolutely necessary, but also because you don't want to contaminate the source; particularly with groundwater. The measures taken to prevent contamination occurring are known as source protection, and in future are likely to be incorporated into regulations on private water supplies (as discussed in Chapter Four).

Springs

Spring water is an excellent source; it will be cleaner than surface water, and, since it reaches the ground surface naturally, you may not need to pump it. If you are accessing a spring, you will need to build a structure around it that protects it from contamination. This is known as a 'spring box', and is illustrated in Figures 2.2 and 2.3. The spring water enters this box via deliberate gaps in the block work, through a gravel base, or via a pipe buried in loose material behind the box. The box allows some storage to buffer variations in demand and protects the spring from contamination. An area extending at least 4 metres around the spring should be fenced off to prevent access by animals. You should dig an interceptor ditch around the uphill side, to channel surface water away from the spring. The spring box can be built out of block work or concrete rings, and must have a secured lid, a vermin proof overflow and a drain as well as the water outlet. The overflow will flow more or less continuously and it is important to allow this to happen. You should not use any pressure within the groundwater to pump the water to height. Both these actions could change the natural hydraulic pressure gradients within the groundwater and potentially cause the eye of the spring to move. It would also provide a contamination route into the groundwater.

Modified spring designs

At some sites there will be a distinctive 'spring line' where water emerges over a more dispersed area at the same height across a hillside. In this instance, a modified structure of a spring box collecting from an infiltration area (Figure 2.4) will maximise the water yield.

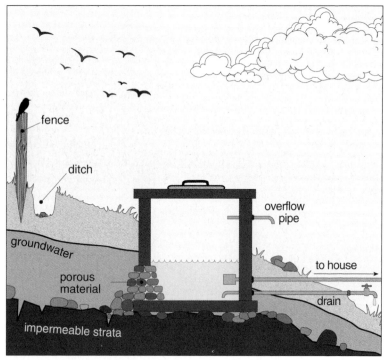

Figure 2.2 Cross-section of a spring box.

At its simplest (e.g. a water supply for an animal drinking trough), a spring can be captured and safeguarded simply by driving a pipe horizontally into the eye of the spring, and fencing around the area to prevent contamination from surface water and animal faeces. It is important that the pipe itself does not allow contamination into the water source (so it must have a trap). This solution is sometimes known as a 'horizontal well' and is commonly used by farmers to convert boggy patches in fields into useful water supplies.

Case study – an isolated cottage in Wales served by a spring

Figure 2.3 shows a spring box serving a 2-bedroom house on a remote Welsh hillside. The nearest water main runs alongside a road a mile and a half away. The cottage is on a steeply sloping site that forms a slight valley in a large grassy field. A conspicuous boggy patch in the field indicated the presence of a spring, fed from the large catchment above. This area has been developed into a spring box which serves the house. It is built of bricks on a slate slab base, and the outlet pipe is 25 millimetre in diameter and is covered by a coarse strainer. The volume of the spring box is about 1.75m^3, and it is situated about 10 metres above the house. There is therefore sufficient water pressure to feed the house without needing a pump. The spring box therefore functions as both a store of water to cope with peak demands, and as a header tank. The residents are water efficient and the spring has never run dry.

Figure 2.3 Spring box, Wales. This spring box serves a 2-bedroom house. There are spaces between the blocks underground on the upslope side, allowing water to enter the spring box. The water pipe leading to the house leaves the spring box at ground level at the front of the structure; the upper pipe is the overflow. The spring box would be improved by a more tightly fitting lid, an interceptor ditch and a fence to prevent access by animals. The overflow pipe overflows to ground, causing the boggy area in front of the spring box.

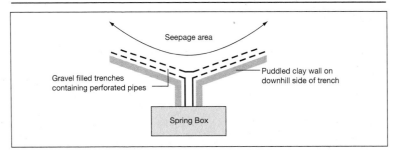

Figure 2.4 A spring with infiltration area (plan view). This technique maximises yield in situations where the spring hasn't got a single 'eye'

Wells and boreholes

If your groundwater supply isn't bubbling to the surface in a spring, you will need to sink a well or a borehole. Wells and boreholes have similar features, but the dimensions are different. A well is generally of a large diameter (at least a metre), dug by hand or mechanical excavator and then lined with concrete rings or masonry. Boreholes are smaller in diameter (6 inches is usual for domestic use) and are excavated, by drilling, and then lined with steel or plastic tubes. If you are accessing water from below 4-5 metres, a borehole is likely to be more economic and easier to build than a well. Both wells and boreholes are classified as 'shallow' or 'deep'. This is not a reference to how far below ground level they extend, but rather to what level of aquifer they are abstracting water from and both are illustrated in Figure 2.1. Shallow wells and boreholes are more likely to dry up (or at least suffer from large changes in water level) than deep ones.

The major features of constructed wells and boreholes are illustrated in Figure 2.5. To prevent contamination from surface water, either the top of the borehole projects above the ground surface, or the area around the top slopes away from the borehole. Contamination may also come from water in the topsoil and subsoil, so the solid lining should extend down the sides of the well/borehole into the aquifer.

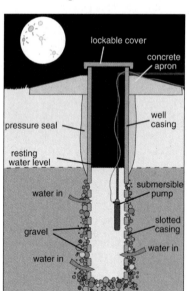

Figure 2.5 Important features of wells and boreholes.

Below this point perforated lining is used. The central shaft provides a buffer, and gives an indication of the level of the water. A dip pipe should be incorporated so that the level can be checked if desired. There may well be gravel both at the base of the well/borehole, and around the perforated pipe, to prevent silt particles entering the shaft. Most people will leave well digging and borehole drilling to a professional contractor (as boreholes are often 60 metres deep, specialist machinery is required) and you should refer to the Environment Agency guide to ensure that all works are carried out according to best practice. Many professionals are members of the Well Drillers Association. If you are interested in digging a well yourself, detailed information is available in *Hand dug wells and their construction,* details of which are listed in Further Reading. It is possible in some instances to combine a private water supply and a closed loop ground source heat pump in the same borehole.

Reinstating old groundwater supplies

If you have an abandoned water supply that you would like to reinstate, the first thing you should do is try and establish why the source was abandoned in the first place, to see if it's worth putting any effort into. Did it run dry, was it of poor quality, or did it simply fall into disrepair when mains water became available? Depending on what you find from these initial enquiries you may choose to unblock the well.

How easy this will be depends upon how it was sealed off in the first place. It may have been simply covered with a slab and then become overgrown, or it may have been filled with rubble, concrete or even rubbish. If the well has been filled in with concrete, it may be easier to dig a new one into the same aquifer than to reinstate the old well. It goes without saying that you should never go down a well to investigate its condition without taking adequate health and safety precautions. Once you have accessed the old supply, you should find out how deep it is and what the water level is. You can check the water level by lowering a 'dipper' into the well. The professional versions consist of two electrodes on the end of a long tape measure; you lower the dipper down the well and when the electrodes hit the water surface a circuit is completed that makes a buzzer go off. You can improvise a dipper using a 250 millilitre bottle with some sand in it (so it floats upright), tied to a measuring tape or piece of string (Figure 2.6). You will be able to 'feel' the water surface by the change in weight on the end of the line. Take a measurement at this point. Next, take something that will sink and attach it to the measuring tape, to determine what the bottom level of the well is. You can then calculate the amount of water in the well.

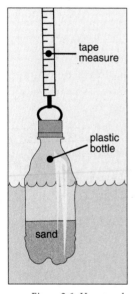

Figure 2.6 Home made dipper. You can test the water level in an existing well using a 'dipper' made out of an old plastic bottle with sand in the bottom, attached to a tape measure. The sand weights the bottle slightly so that when you lower it into a well, you can feel the change when the bottle hits the water.

Get any water that's in the well pumped out; if the well hasn't been used for years then the water it contains will not be representative of the water that will recharge it. The next thing to do is to see if the well refills once it has been pumped out. If it doesn't, you should examine the lining of the well at the bottom; it may have collapsed, or become blocked up with silt so water can't enter it any more. There may be chemical precipitation or bacterial slime on the lining that has decreased its permeability. These problems can be cured (by mechanical scrubbing or by acid cleaning), but these are jobs you should leave to specialists. If having checked these possibilities the well still doesn't refill, then it's probably not worth going to any more trouble and you should look at digging a new well or borehole.

Assuming the bottom of the well has refilled, you should get the water analysed to see what's in it. Even if you want it for garden watering rather than for drinking you will want to check that it hasn't got high levels of metals or salts in it. Once you have determined that the quality is suitable, you will ascertain the available quantity. This will require a pumping test, as described in the box on page 38. Having been through the process of discovering what the water quality is and ascertaining that there is enough of it, you should undertake a thorough examination of the lining and superstructure of the well, and either upgrade or rebuild, incorporating the features illustrated in Figure 2.5.

Streams or rivers

How you choose to take the water out of a stream or river depends primarily on how much the water level changes and what the water quality is like. There are three main methods, of varying degrees of sophistication.

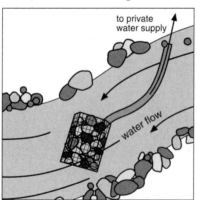

a) Simple inlet pipes

If you just put the end of your pipe in a stream and weigh it down, it may well get blocked fairly quickly. A simple strainer, as shown in Figure 2.7, will reduce this risk. A strainer with a rounded or pointed end will be less liable to blockage. Try and position the inlet in a straight part of the stream to minimise the amount of solids that will get swept in by turbulence.

Figure 2.7 An inlet pipe with a rock-filled basket surrounding it to provide some coarse filtration.

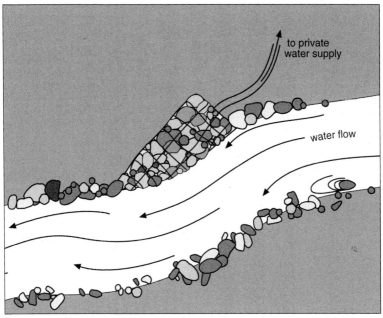

Figure 2.8 Inlet built into the bank of a stream. A stone-filled wire basket (known as a gabion) is buried in the bank, with the end of the inlet pipe inside it. This acts as a filter to prevent debris entering your water supply.

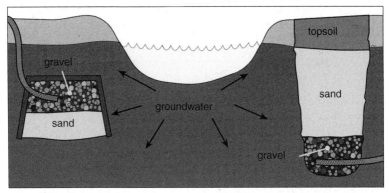

Figure 2.9 Infiltration galleries. When a stream or river is a result of the water table reaching the surface, the banks of the watercourse will be permeable, allowing water to be naturally filtered through the bank before it enters your water supply. The system on the left consists of an upside down tank, which can either be buried in the bottom of the stream, or to one side of it. The system on the right could be left open above ground as a well. In both cases you will need to divert the water during construction.

45

b) Inlets built into the bank.

Excavating part of the bank (as shown in Figure 2.8) so that the inlet pipe faces downstream will reduce the amount of solids entering the pipe. You should reinforce the bank around the inlet with gabions (stone-filled wire baskets) or large rocks.

c) Infiltration galleries

These are more sophisticated strainers (Figure 2.9), sometimes used when the river contains a lot of silt. An upturned tank (e.g. a domestic header tank) containing gravel can be buried next to the stream, or under the stream-bed if it is sandy and you can build a temporary dam to keep the area dry whilst you are working. Choice of location for these structures is critical; the stream-bed may move, and the gravel in the infiltration gallery may regularly become blocked with silt from the sand bed. In most instances, this is an overcomplicated solution that does not justify the difficulty of installation.

A variation on an infiltration gallery is to dig a well adjacent to the surface water supply and use the natural filtering properties of the river-bank to remove much of the suspended solids that are contained in surface water sources. The void of the well also provides some storage capacity.

Case study – The Centre for Alternative Technology

CAT is situated on the site of an abandoned slate quarry in mid-Wales. In the middle of the 1800's, the slate miners built a reservoir, dammed with slate and soil, so they could drive a water wheel, which in turn powered slate cutting machinery. The reservoir has a capacity of around 5 million litres, and is fed by two streams, which are fed from rainfall on unimproved upland grazing land. Water exits the reservoir via a 200mm diameter pipe, which runs across the hillside for 200 metres before the drinking water is 'T'-sectioned off in a 63 milimetre pipe. This water passes through a slow sand filter (discussed in Chapter Four) and into storage tanks (with a total volume of around $7m^3$) that provide a buffer to cope with peak water demands. It then flows down a 63 milimetre pipe to the level of the site (some 30 metres below), where it passes through a manifold containing 4 ultraviolet bulbs that kill any remaining pathogens in the water (UV for water treatment is discussed in Chapter Four). The remaining water, not used for domestic purposes, continues along the hillside in the 200 milimetre pipe before dropping down to the level of the main site through a hydroelectric turbine and into another reservoir. From this point the water either flows down through a second hydroelectric turbine 30 metres below, or is used to drive a water-balanced cliff railway.

Existing lakes and reservoirs

Existing lakes and reservoirs can be good sources of water for private supply, providing they are large enough. You will need to monitor the water level over time to see how much it changes. You should find out how and where the lake fills and empties from; in some instances you could be better off collecting from one of these sources rather than from the lake itself. Is there much surface water run off into the lake that will affect water quality? Outlet structures should take advantage of natural settlement and can consist of an outlet pipe with a strainer held above the lake sediment by a float and anchor (made on a similar principle to that shown in Figure 5.7, but with the addition of an anchor and a more robust float).

If there isn't enough in the stream – building a reservoir
If the flow in a surface water source is too variable to be a reliable supply, you may be considering building a reservoir. Since this can seriously disrupt the aquatic life in the water source, the Environment Agency must be consulted and they will probably limit the amount of water you are allowed to remove. This will usually be based on maintaining dry weather flow in the watercourse, so you must ensure that the excess that you can abstract is sufficient for your needs. Building reservoirs is best left to a professional. They will calculate the necessary size, carry out the excavations and install the drainage system for you. Reservoirs are expensive to build, particularly if the local soil is not sufficiently impermeable to provide a lining and build the dam with. There are two basic types; those that involve damming a watercourse (impounding reservoirs) and those filled by the stream but dug adjacent to it (off-stream reservoirs).

Source protection

When dealing with private water supplies, you should do all you can to protect your water source. Specific source protection mechanisms for the various structures you may build around your water source are described above, but the following general points should be borne in mind.

Actively minimise water use

Depending on the nature of the source, there may be a risk of it running dry, this will be obvious with surface water, but you may have little warning that this is going to occur with groundwater sources. This is not just a problem of insufficient quantity available to you. In the process of running dry, the water quality may decrease considerably (e.g. as you start using more stagnant water from the bottom of a lake), which may mean it needs more treatment. Running it dry will also seriously damage the ecosystem from which you are taking the water: if there is a risk of this occurring the authorities will have put limits on the amount of water you are allowed to abstract. Even if it appears plentiful, it is good practice to be as economical as possible with your water, as described in Chapter One. You should also be aware of who else is abstracting water from the same source as you, and what the consequences of level changes are to them; depending on the local geology, they may run out of water before you do.

Not making what you are about to use dirty

If you are going to a lot of trouble to find your water source, it clearly makes sense to try and keep it clean whilst you are abstracting it and moving it to where you want it.

Not contaminating your source

Whatever structures you put in or near the water source to enable you to collect it, you must make sure that they are clean, and are certified for use with potable water sources. Source contamination is a particular concern with groundwater supplies. If you are opening up a groundwater supply to the surface that was previously sealed, you could potentially pollute an entire aquifer, for example with the pathogenic organisms found in animal faeces that may be present in surface water run off, or from chemicals leaching out of your pipe or water tank. You must also prevent the backflow of water into the groundwater source (since it may have been contaminated whilst above ground level), and for the same reason you cannot put water down boreholes or into groundwater without licencing and adequate treatment.

Summary

Your main concerns when choosing a private water supply will be the quantity and quality of the water available to you. Your site may have a number of sources available, in which case the following is offered as guidance on which sources are most likely to be appropriate:

1) Mains supply – clean, plentiful and cheap!

2) Spring

3) Well or borehole

4) Surface water

5) Direct rain – considered in Chapter Five

6) Reused water – considered in Chapter Six

Further Reading

• *Oxfam Water Supply Scheme for Emergencies*. Instruction manual for Hand Dug Well Equipment. Available to download from the Oxfam website (www. oxfam.org.uk) – designed for field engineers using Oxfam emergency kits, but provides useful detail on wells and boreholes

- *Going with the Flow*: *Small scale water power*, Langley, Curtis, CAT Publications, 2004. Predominantly a guide to small scale hydroelectric schemes, but has a good chapter on surveying sites for water. Available from CAT Mail Order
- *Field Hydrogeology*, Brassington, Wiley, 1998. Details on site assessment, pump tests for estimating yields of boreholes
- *Hand dug wells and their construction*, Watt & Wood, ITDG, 1979. Useful guide if you intend to dig a well yourself. Available from CAT Mail Order
- Environment Agency Water Resources Officers. Local officers deal with licensing private supplies and are valuable sources of information on local hydrology. They produce a number of leaflets on good practice when constructing wells and boreholes. Local contact details available from www.environment-agency.gov.uk
- Meteorological Office. Has data on rainfall and other weather conditions across the UK. Website at www.met-office.gov.uk
- Centre for Ecology and Hydrology. The main research organisation in hydrology. Website at www.ceh.ac.uk

Chapter 3. Fitting It All Together

Storing water – moving water – the physics of flow – choosing pipes – choosing a pump – wind pumps – hydraulic rams – human powered pumps – electrical pumps – useful controls.

Once you have found your water source, worked out how to get the water out and how to clean it, you need to think about getting it from the source to your house. You will want to do this in the most energy efficient way possible and without contaminating the water. You will need to choose pipes and fittings, and, if you need to move the water uphill or over some distance, you will need to choose a pump. Your water system will have the following components:

• an inlet structure (considered in Chapter Two)
• storage
• pipes
• an energy source
• control mechanisms
• cleaning processes (considered in Chapter Four)

The precise layout will vary between systems, particularly in terms of the location of the treatment and storage components.

Water storage

Water storage is needed to buffer differences between supply and demand, and to provide a backup in the case of failure of the source or pump. In some cases the storage also provides some treatment (e.g. by settlement). The two main types of storage tank (or cistern) are open tanks and pressure tanks.

What type of storage tank should I choose?

Open storage tanks (Figure 3.1) are so called because they are open to atmospheric pressure. It is necessary to incorporate an overflow in your storage tank; whilst this may seem like a waste of water, it is important that you don't allow your stored water to stagnate over a period of weeks. You may be able to put this overflow water to good use, or return it to the water source (e.g. surface water) so that it continues through the hydrological cycle. The most common materials for potable water tanks are plastic and metal, and both types are available in a variety of shapes and sizes. When buying tanks you should ensure that they are suitable for potable water and, if they are going to be above ground, that the plastic is UV stablised.

Open storage tanks are by far the most common, but if your system is going to include a pump, it may be more energy efficient to have a pressure storage tank. These are similar to expansion vessels in heating systems (although you will need to make sure that you use one suitable for drinking water systems). When electrical pumps start up, they use more energy than they do during normal operation (up to 3 or 4 times as much, depending on the pump type), and they will also wear out faster if they turn on and off frequently. In a private water supply, the pump might therefore turn on every time someone uses even a very small amount of water (e.g. for hand washing), which would be inefficient. One type of pressure tank (a bladder pressure tank) is illustrated in Figure 3.2. Water is pumped into a butyl rubber bladder inside the tank. The rest of the tank contains nitrogen gas. Water entering the bladder causes the bladder to expand, compressing the nitrogen, and the pressure inside the tank rises. When it hits a predetermined pressure level, the pump switches off via a pressure control switch. When a water-using appliance is turned on, water leaves the bladder, the nitrogen expands and the pressure in the tank falls. When the pressure falls to a predetermined level, the pump switches on again and the tank refills. If this pressure range is fairly wide, small amounts of water can be used without the pump being activated, thereby improving efficiency.

An alternative way of preventing the pump turning on and off frequently is to use an electronic pump controller. These consist of a pressure switch and a memory chip that allows the system to work out when to switch the pump on, so as to minimise short pumping durations.

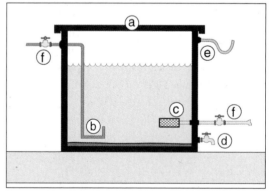

Figure 3.1 Water storage tank.
(a) Lightproof lockable lid
(b) smoothing inlet to maximise settlement
(c) outlet with strainer
(d) drain (e) vermin-proof overflow
(f) stop tap.

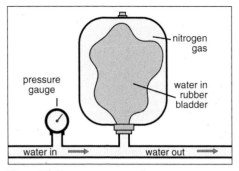

Figure 3.2 Pressure storage tank.

How much to store

There are two issues to consider with regard to how much water to store. Firstly, your total water use and, secondly, peak demands. In an open storage tank, you should try to store enough water for around one to two days' use (you should have a good idea of how to calculate this from Chapter One). This allows peak water demand to be met (e.g. filling a bath), but will allow the storage tank to overflow at other times (so that it doesn't become stagnant). If there is a chance that your water supply will dry up for a period, you will need to store enough to last for this time. Ultra low water use households may have peaks in demand that easily double average daily use (such as for an occasional bath instead of a shower). If you have regular guests, you should take these peak demands into account when deciding on a size of storage tank.

If your water use varies according to the season, you may choose to have reserve storage tanks that you only use for part of the year, although you should make sure that these are cleaned and disinfected before being brought into use. If this extra demand is for non-potable water, such as garden use, consider having a separate store of non-potable water for this purpose.

If you are abstracting groundwater from a spring or a well, you may not need to include a storage tank, since these methods of accessing groundwater often include a storage reservoir; a household spring box can be built to the required size, or if you have a well, the cavity within it will act as storage. However, if you are pumping water from your groundwater source rather than relying on gravity, incorporating storage will increase the efficiency of the pump (as discussed above), and provide a buffer in the event of pump failure.

How to move the water – choosing an energy source

How much energy do you need?

Moving water takes energy. If your water source is significantly higher than your house, you may not need to add any extra energy; water will flow from a high point to a low point under gravity. As well as the height difference, you will also need to consider what water pressure and flow rate you need within the house, and the highest point in the house at which you will need these water pressures. You may also find that there are other components of your system that cause drops in pressure, and you will need to take these into account when working out how best to move the water.

1) Measuring the height difference

You will need to measure the difference in height between your source and your supply point. This is known as the static head, is usually expressed in metres, and is the major determinant of energy requirements. When you measure this difference you should determine the level of a few landmarks along the way as well as the total head so that you can create a profile, which will allow you to choose an optimum position for your storage and treatment systems. There are several ways of measuring differences in height, two of which are described on page 55 opposite.

2) Pressures, flow rates and head requirements within the house

If you have a mains water supply, your water will be at around 2-3 bars of pressure (although this will vary according to the demands on the network at a given moment). Private water supplies may not need to run at pressures this high, and your pressure requirements will vary according to your appliances. Many point-of-use water treatment systems will have minimum operating pressures (typically 0.5 bar for under-sink filter units). An instantaneously heated electric shower will usually have a requirement for a minimum of 1 bar (but, given their energy and water inefficiency, you should try and avoid these anyway!). If your hot water system includes a header tank, you will want the water to fill the tank at a sufficient flow rate to allow you to run a bath or have a shower without the water stopping because the header tank has emptied. You could have a large header tank (sized according to your maximum hot water use, e.g. 100 litres for a bath), so that you don't run out of water even if the fill rate of the header tank is slow. In this case, you will simply need to make sure that you take the height of the header tank and any pipe losses into account when calculating your total head. Alternatively, you could have a smaller header tank but use a higher water pressure so it fills more quickly.

In practice, your choices about pressure and flow will probably depend upon whether or not you are expecting to need a pump. If you are operating a gravity fed system, you should ensure that there is enough head to fill your header tank, and increase the size of the tank if necessary to compensate for a slow fill. If, based on your initial head calculations you are expecting to need a pump, then you should check with the pump supplier what pressures the pump will generate (although you

Figure 3.3 'Dumpy' laser level used to measure the height difference between two points.

Measuring height differences

A laser level (Figure 3.3) consists of a tripod stand with a laser light on the top that you project horizontally onto a staff held further up the slope. The staff has markings on it so that you can record the height difference between the laser and the staff. This is repeated as many times as necessary until you get to the top of the slope. They are widely available, so you should be able to hire or borrow one.

A clinometer (Figure 3.5) measures angles, and some basic trigonometry will enable you to turn an angle into a height difference. Take two posts of the same height and position one at the bottom of the slope and the other further up (as far away as you can comfortably see it). Place the clinometer on top of the first post and sight to the top of the second post, taking a note of the angle. Record the distance along the ground between the two posts, as well. Repeat this procedure by moving the posts until you get to the top of the slope. You can then use the sine rule to calculate the vertical distance between points (Figure 3.4). Clinometers can be bought from navigation or geological suppliers and are much quicker for measuring big differences in height than laser levels. They can also measure angles below the horizontal, so are easier to use on undulating ground than laser levels. However, they require you to be able to accurately measure the distance along the ground.

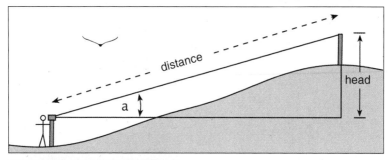

Figure 3.5 Clinometer, used to measure gradients.

Figure 3.4 Measuring head using a clinometer and the sine rule. You will know the angle of the slope from the clinometer. The sine of this angle, multiplied by the distance between the two points along the ground will equal the head. For example, if the clinometer reading is 14°, and the distance along the ground is 35 metres, the head is sine 14° x 35 = 8.5 metres.

can generally assume that even the smallest pump will provide water at over a bar of pressure).

3) Other sources of head loss

The total head which we discussed measuring on page 54 will not necessarily be the difference in height between your water source and the points of water use in the house. If you have any storage or treatment components within the system that are at atmospheric pressure, it is this point that will be relevant in the head calculation, as shown in Figure 3.6. You must take this into account when positioning treatment and storage tanks. You would place them as high up in the system as possible if short of pressure and head. In the unlikely event of the water pressure being too high due to a high head, a storage tank can be included specifically to reduce pressure (in which case it is known as a break-pressure tank). However, a pressure-reducing valve or a flow restrictor may be a more straightforward solution.

When water in the system isn't moving, a pressure gauge will give a reading that is referred to as static pressure. If you then turn on a tap, the pressure will drop. The lower pressure that occurs when the water is flowing is termed dynamic pressure and it is this pressure that will determine how fast water will come out of a tap. The reason that the pressure drops is that energy is lost in the pipe and fittings. The energy lost within the pipe is dependent on the pipe diameter and the flow rate, and can be calculated as discussed in the box on head loss. The amount

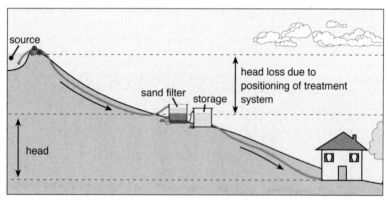

Figure 3.6 Head loss due to treatment system. In this example, the actual head of pressure available in the house is lower than the head difference between the source and the house, since the treatment system is positioned part of the way down the hill and is open to atmospheric pressure. To maximise the available head, the treatment system could be moved further up the hill. Within the house itself, the height above ground floor level that you have water using appliances will also need to be taken into account (but is not illustrated here).

of energy lost within bends and pipe fittings varies according to the type of fitting, and can be expressed as an equivalent length of straight pipe and then added to the losses you calculated for the pipe itself. An example of this calculation process is described in the box on page 58, along with some more of the physics that governs pressure and flow.

Choosing pipes

Water pipes suitable for carrying domestic water supplies come in a variety of materials, the most common of which are shown in Figure 3.7 and compared in Table 3.1.

Polythene MDPE pipe is by far the most common (and environmental) choice to move water between the source and the house, with copper or polybutylene used within the house, as they are compatible with hot water systems. Multilayer composite pipe (aluminium and plastic) is occasionally used for water mains where the ground is contaminated. Whatever sort of pipe you use, check that it is suitable for potable water supplies so you don't get chemicals leaching into your water supply, and that the rated pressure is sufficient (for a private water supply it is unlikely that you will need anything other than the standard rating for MDPE which is 12 bar). You should make sure you have your water tested (as described in Chapter Four), as corrosive water may influence your choice of pipe material. The type of plumbing fittings you use will be also be determined by your choice of pipe.

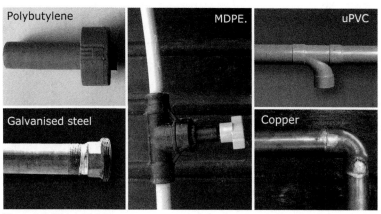

Figure 3.7 Pipe materials.

Material	Comments
Polythene (polyethylene)	Flexible, rot resistant and fairly robust. Sold in coils up to 150 metres. If subjected to frost it will tend to burst at the joints rather than split the pipe, making repairs much easier than with copper. Two types are available; medium-density polythene (MDPE) and cross-linked polythene (PEX). Not suited to hot water. Standard sizes are 20, 25, 32, 40, 50, 63mm. Usually blue, (old pipe is black).
Galvanised steel	Not recommended owing to corrosion problems.
uPVC	More commonly used for foul water drainage than potable water supply.
Polybutylene (PB)	Flexible and robust, can be used for hot water (but not within a metre of the boiler). Commonly used within the house.
Copper	Maintains shape; can be bent if a pipe bender is used to maintain the diameter, expensive for long supply pipe runs with large diameter but is extremely suited to shorter pipe runs in houses. Sold in 3 metre lengths. Connected by soldered joints (cheap) or compression fittings (more expensive). Will split if frozen.

Table 3.1 Common pipe materials and their features.

Calculating head loss in water supply systems

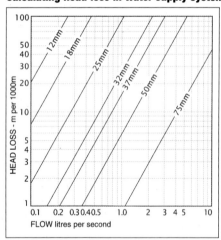

Figure 3.8 This graph shows the amount of energy (expressed as head) lost for every kilometre of MDPE pipe. This varies according to the flow and pipe diameter. Similar graphs are available for pipes of different materials (courtesy of Intermediate Technology Development Group).

As discussed earlier, head is the vertical distance between two points; for example your water source and your house, measured in metres. Pressure is the force that something exerts, and is proportional to the density of the water (which is constant) and the head. Pressure can be measured in Pascals, bars, or pounds per square inch. The static pressure in a water system is the force that the water exerts when it is not moving, for example when there are no taps turned on. If a tap is turned on, water begins to flow down the pipe. As it does so, energy is lost due to friction against pipe walls and turbulence caused by bends and pipe fittings. The amount of energy lost varies according to the flow rate of the water. The amount of pressure that is left after these losses is known as the dynamic pressure. It is this dynamic pressure that determines the water flow through appliances within your home. The difference between static

Fitting	Equivalent Loss (head, in metres)
90° elbow	15 x pipe diameter
Tee (perpendicular to flow)	15 x pipe diameter
Tee (aligned with flow)	6 x pipe diameter
Foot valve	40 x pipe diameter
Gate valve	3 x pipe diameter

Table 3.2 Energy loss due to pipe fittings (assuming a flow rate of 1.5 litres/second).

pressure and dynamic pressure is sometimes referred to as head loss; the height of water that is required to overcome the pressure loss. Most of the time, we are more interested in dynamic pressure than in static pressure, so when a pressure is measured it is often quoted as a value at a certain flow rate. You can calculate the losses due to friction and turbulence, and you will need to do this if you are going to specify a pump, or determine whether or not water will flow through your system under gravity alone.

The amount of friction is dependent on the smoothness of the pipe wall, and the speed at which the water is flowing along the pipe and varies according to pipe material and size. The graph in Figure 3.8 shows the energy loss (expressed as vertical head of water) for MDPE pipes of different diameters at different flow rates. Similar graphs or tables exist for all pipe types. There will also be losses in the pipe due to turbulence caused by obstructions such as bends and taps. Table 3.2 shows an equivalent length of pipe so that you calculate these losses as well. If you don't want to do the maths, then you can assume that 25 milimetre MDPE will be sufficient for single house applications where the supply pipe is less than about 300 metres long; head losses are low unless you are transporting water large distances, only have a very small pipe, or need a particularly high flow rate (e.g. for power showers).

Note: The mathematical relationship between water pressure and flow is given by Bernoulli's equation and is beyond the scope of this book. It can be used to calculate water pressure under dynamic conditions, i.e. when water is flowing at a known rate.

Laying pipe

Your water pipe will need to be buried in a trench to protect it from mechanical damage, ultraviolet light and frost. Depths are specified in the Water Regulations: 50 centimetres is common, deeper if heavy vehicles may be passing overhead. Given the labour required for this, you may prefer to run the risk of occasional freezing. You should make sure when laying the pipe that it is free from kinks, sudden changes in direction or height, and that there isn't anything sharp in the trench that could damage it. Don't pull the pipe completely straight; small bends should be left to allow movement should ground settlement occur. The system should be checked for water tightness before backfilling the trench, and you should avoid burying any connections or taps; install a manhole at these points. At points where the pipe is going to be exposed to the air, it should be lagged to prevent freezing. The point at which freezing is most likely to occur is around any float valves, so pay particular attention to them when lagging.

Head	Capacity	
(metres)	m³/hour	Litres/sec
10	4.8	1.25
15	4.2	1.17
20	3.6	1
25	2.9	0.8
30	1.8	0.5
35	0.6	0.17

Table 3.3 Pump output versus head for a typical submersible pump suited to domestic supplies.

Pumps

By this stage, you will have got as far as choosing a type of pipe and how big it should be. You will also have measured the head in your system, and calculated head losses due to the pipe and fittings. The end result is the total head within the system. You can then take this value, and the amount of water you need to move (which you worked out in Chapter One) so that you can work out what energy source is most appropriate for your water system. If the head in your system is sufficient to produce an adequate flow rate into your header tank and to all your water using appliances, you may not need to pump your water at all. However, many instances will require some sort of pump, and the following is designed to help you choose what sort of pump is most appropriate. Principles and mechanisms of pumping are dealt with in the box below. Pumps are specified by referring to their output (in litres or cubic metres per unit time) versus the head, and this is either given in a graph or a table, as shown in Table 3.3 (for a typical submersible pump for small scale applications).

How pumps work

The action of a pump is of two sorts. Firstly, the pump must 'pull' water into itself and secondly, it must push water out of the other side. The relative amount of each of these actions leads to classification of pumps into those that push water (*submersible pumps*) and those that pull water (*suction pumps*), i.e. pumps that are situated above the water (suction pumps) and those in the water (submersible pumps). The practical difference between the two types is the amount of head they can pump water to. Suction pumps cannot move water to a head higher than around 8 metres (they are limited by atmospheric pressure, so the theoretical maximum pumping head is 10.4 metres at sea level, but is lower in practice due to efficiency losses), whereas submersible pumps can be found operating in boreholes that are 70 metres deep. Suction pumps will not operate if the supply pipe contains air; the pipe needs to be filled with water in a process called priming. Some suction pumps are self-priming; they can remove small amounts of air that may enter the pump during operation, but you will generally need to supply the initial priming water.
Pumps can also be defined according to their mechanisms of action:

Piston pumps: the classical design used for hand operated pumps. On one stroke a plunger creates suction that draws water through a valve into a cylinder. On the return stroke the water is forced out through a second valve. If pumping against a large head, the plunger needs to be at depth so that most of the action is via pushing rather than pulling.

Rotary, centrifugal and helical pumps: these have a rotating part called the impeller (like a propeller), which as it rotates tends to suck water into the pump and push it out of the other side. In helical rotor pumps, the impeller consists of a spiral shaped rod. Centrifugal and helical pumps can operate to considerable heads, and can be combined in a multi-stage pump in high head situations (each stage is basically a pump with its own impeller, that pushes water up to the height of the next pump).

Jet pumps: Water travels through a narrow tube, which widens at the outlet end. As water flows into the wider section, it creates a partial vacuum which sucks more water in to the pump.

Peristaltic pumps: a flexible tube is squeezed by rollers, thereby pushing the liquid through the pipe. They are not used for water supply pumps.

Figure 3.9 Windpumps can be used to abstract water from wells and boreholes. The spinning blades are attached to a crank mechanism which drives a piston pump. Windpumps can move thousands of litres a day from depths of over 50 metres, with a zero electricity cost, but their capital cost is high and so they are best suited to large supplies or where electricity is not available.

There are four main energy sources available to you that could power a pump; wind, water, human and electrical.

Wind and water powered pumps

From the point of view of sustainability, you should try to minimise your electricity use. In certain circumstances, both wind and water can be used to power pumps directly, without converting the energy to electrical power first.

Wind pumps (Figure 3.9) have been used for hundreds of years to pump water from wells and boreholes. They consist of a multi-bladed rotor, attached to a crank acting on a reciprocating piston. Of course, they only pump when the wind is blowing, so you will either need to store enough water to see you through calm periods, or have a standby pump powered by another means. An additional problem is that the mechanism will be damaged if you try to prevent the pump from pumping water when it is windy, so you will need to incorporate a facility for overflow from the storage tank, although preferably not one that will be wasted (you should not connect the overflow back into the well/borehole directly as it will potentially have become contaminated by reaching the ground surface). The capital cost of wind pumps is high (over £2000), so they will only be economic for large scale supplies where running costs would outweigh capital costs (i.e. thousands of litres/day), or where there is no mains electricity supply available. In these instances, the wind resource should be estimated and used to specify the correct capacity of wind pump and a suitable storage volume (see Further Reading for details).

Hydraulic ram pumps

Hydraulic rams (Figure 3.10) are water powered pumps that use the kinetic energy in the incoming water to alternately open two valves and pump a proportion of the incoming water to a height (see Figure 3.11 for a detailed explanation of how they work). The proportion pumped to this height depends upon the head available in the incoming water, and the head to which the water must be pumped (Table 3.4). They can therefore work very well in situations where surface water from a stream is being pumped to some distance above it.

Figure 3.10 (right) Hydraulic ram pump.
Figure 3.11 (below) Hydraulic ram operation
a) pulse valve, b) delivery valve, c) delivery chamber,
d) supply pipe, e) drive pipe. 1) Water flows into the
hydraulic ram via the drive pipe (e)
2) When the pulse valve (a) snaps shut it causes a
pressure wave that forces the delivery valve (b) open.
Water then enters the delivery chamber (c) and leaves
through the supply pipe (d).
3) When the pressure in the supply chamber falls below
that in the delivery chamber, the delivery valve snaps
shut and a pressure wave is transmitted back through
the supply chamber, opening the pulse valve and
restarting the cycle.

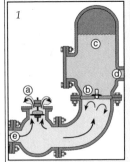 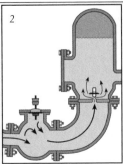 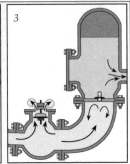

Case study – remote holiday cottages using a hydraulic ram pump

The conversion of a collection of barns into holiday cottages, on a remote Welsh hillside, required the installation of a water supply to serve up to 10 people. A stream runs past above the house, about 100 metres away, but the water would have required pumping to boost pressure through the treatment system. There were additional concerns over land ownership, the presence of a bridleway and access agreements. There is a larger stream about 10 vertical metres below the house, with no access or ownership concerns. A 1.5 inch diameter hydraulic ram with a drive head of 5 metres pumps water up to 30 metres above the house where there is a slow sand filter and storage tank. Water then flows by gravity to the holiday cottages, which have a UV filter to kill micro-organisms.

Working fall or head (metres)	Vertical height to which water is raised above the hydraulic ram (metres)											
	5	7.5	10	15	20	30	40	50	60	80	100	125
1	144	77	65	33	29	19.5	12.5					
1.5		135	96.5	70	54	36	19	15				
2		220	156	105	79	53	33	25	19.5	12.5		
2.5		280	200	125	100	66	40.5	32.5	24	15.5	12	
3			260	180	130	87	65	51	40	27	17.5	12
3.5				215	150	100	75	60	46	31.5	20	14
4				255	173	115	86	69	53	36	23	16
5				310	236	155	118	94	71.5	50	36	23
6					282	185	140	112	93.5	64.5	47.5	34.5
7						216	163	130	109	82	60	48
8							187	149	125	94	69	55
9							212	168	140	105	84	62
10	Litres pumped in 24 hours per litre/min of drive water					245	187	156	117	93	69	
12						295	225	187	140	113	83	
14							265	218	167	132	97	
16								250	187	150	110	
18								280	210	169	124	
20									237	188	140	

Table 3.4 Hydraulic ram output (courtesy of Allspeeds Ltd.).

Human powered pumps

Piston pumps can be hand powered, with the delivery rate varying according to the depth of the well/borehole. They are commonly specified for developing-nation water supply projects but there is no reason why you shouldn't use one for your private water supply. You should ensure that you get one that is genuinely designed for use (e.g. those used by development agencies, Figure 3.12), rather than those designed to look nice in ornamental gardens. A hand pump might also be an appropriate backup system if you are installing a wind pump. It is important to specify the hand pump to the depth of the well; as discussed in the box on pages 60-61, a hand pump that is relying on suction can only be used in shallow wells (where the water level is less than 7 metres below the ground surface). Hand pumps for deeper wells and boreholes must have a slightly different configuration, so that the action is more that of a submersible pump than a suction pump. Water drawn by hand pumps should be fed directly to storage tanks to avoid contamination; if these tanks are significantly above the top of the well/borehole, a 'force pump' will be required (a modified type of piston pump).

Hand-operated diaphragm pumps (also known as bilge pumps, Figure 3.13) are very simple, and can pump to heads of around 3 metres and 40 litres/minute,

Figure 3.12 (left) A robust piston driven hand pump (Nira AF-85).
Figure 3.13 (above) A bilge pump has a rubber diaphragm in it that is moved up and down with the handle. This alternately creates suction and positive pressure, which drives water through the pump owing to the valves at either end which prevent the water flowing in the wrong direction.

and it may be worth having one of these pumps for use in an emergency if your main pump fails. The more robust versions available from marine stores are suited to long-term use if you wish to pump all of your water by hand, and they are considerably cheaper than piston hand pumps (although since they are suction pumps they are not suited to deep well or borehole applications).

Electrical pumps

A wide variety of electrically powered submersible and suction pumps is available. Advantages and disadvantages of each are described in Table 3.5, and their appropriateness for various situations is described below.

Groundwater sources

If you are having a well or borehole dug or drilled, the contractor will specify and install a pump for you. Submersible borehole pumps as shown in Figure 3.14 are specifically designed for the purpose, and can be specified to pump to heads of over 100 metres if necessary. They are generally centrifugal pumps and may consist of several 'stages' if specified for deeper boreholes. If you have a shallow well or borehole (less than about 5 metres) you could in theory use a suction pump (Figure 3.15), although since they are less efficient, this is not usually recommended.

Surface water sources

The type of pump best suited to surface water sources depends on the inlet structure you are using to abstract water. Submersible pumps should not be placed directly into flowing watercourses where they are likely to get damaged by silt and debris, but if the inlet structure includes an infiltration gallery, or consists of a well or sump dug next to the river bank, then a submersible pump is possible. If you are using a very simple inlet structure then you will need either to gravity feed the water to a tank that can contain a submersible pump, or use a suction pump located on the bank in a pump house.

The type of pump used also depends on the type of treatment system you are using (which, as you will see in Chapter Four, depends on the contaminants in the source water). Generally, if you need to generate water pressure for the treatment system (e.g. to pass the water through a filter), you will need to position a submersible pump so that it is pushing water through the treatment system, or pump to a height above it so that water can flow down through the treatment process by gravity.

Photovoltaic powered pumps

Most electrical pumps run from a mains electricity supply (AC current). If you wish to power your pump from a photovoltaic panel using DC current directly, you will need a pump specifically designed for the job, since the voltage produced by the

Submersible pumps	Suction pumps
Advantages	Advantages
Do not require priming More efficient Can pump from greater depth	Simple installation
Disadvantages	Disadvantages
May be difficult to remove for maintenance	Will not start automatically if priming water is lost
Installation more complex	Limited head – practical maximum 8 metres. Many models are inefficient beyond 3 metre head

Table 3.5 Comparison of submersible and suction pumps.

panel will vary and this would tend to damage conventional pumps. Accurate site data on the solar resource available is also necessary, and the interested reader is referred to *Tapping the Sun* or *An Introduction to Photovoltaic Power* for further information (see Further Reading).

Controls for electrical pumps

As mentioned above in reference to wind and water powered pumps, one of their disadvantages is that they are difficult to control (i.e. turn on and off). Electrical pumps for domestic water supplies are usually turned on and off by switches designed to sense changes in water level, or in water pressure. Changes in water level are sensed by float switches. These contain a lever operating a switch and a metal ball. When the ball falls it hits the lever and switches the pump on or off (depending on which way round the switch is wired). The location of the switch then determines its function. Float switches located in the water source can be used to switch the pump off to protect it against dry-running. Float switches located in the water storage tank can be used to switch a pump on when the level falls. Modern pressure switches contain a material that changes its electrical resistance according to changes in pressure, although if the pressure switch also incorporates a pressure gauge, it may be an older design that contains a moving diaphragm (similar to that in a traditional barometer).

Pumps will generally come supplied with some kind of strainer over the inlet to prevent solids entering the pump and damaging it. Strainers with floats attached to them are also available (Figure 5.7, page 120), or you can make your own. These must be regularly cleaned so that efficiency is maintained and the pump doesn't get damaged.

Figure 3.14 (left) Submersible borehole pump (courtesy of Amos pumps).
Figure 3.15 (above) Suction pump (Aspira) with pressure switch (courtesy of the Green Shop).

Submersible pumps will have a float switch attached that prevents them from operating when there is no water. It is vital that they do this, since cooling for the motor is generally provided by the water passing through the pump, and if they are allowed to operate dry, the motor will burn out. Suction pumps may also have mechanisms to prevent them from dry-running, and you will need to set up a system to ensure that the supply pipe always contains water. A foot valve (described on page 69) on the supply pipe will help ensure this, but you will need to prime the pump initially (this is done by backfilling the supply pipe with water).

As well as these mechanisms to protect the pump from dry running, you will need to control the pump according to your water use. There are a number of ways to do this, and you can get advice from your installer or the pump supplier. The following are examples of a few potential scenarios.

Example 1 – water from a borehole

Description: the borehole contains a submersible pump. This pumps water to a storage tank, which then gravity feeds the water to the house via a treatment system.

Controls: there is a float switch on the submersible pump to prevent it from dry-running in the event of the level in the well/borehole being too low. The level of water in the storage tank (which could be the header tank in the house) is controlled by a ball valve. When water is used and the ball valve drops, the pressure in the pipe upstream of the storage tank drops, which makes the submersible pump turn on via a pressure switch. An alternative control is to have a float switch in the storage tank that signals to the borehole pump when the water level falls.

Example 2 – surface water supply with suction pump

Description: water is taken from a stream which has been dammed to provide a small reservoir. Located on the bank is a suction pump that removes water from the reservoir and pumps it to a storage tank, which then gravity feeds the water to the house via a treatment system.

Controls: a foot valve on the end of the suction line prevents water draining out of the inlet to the pump. A strainer is incorporated into the suction line to prevent damage to the pump. A float switch in the storage tank switches the pump on and off according to the water demand.

Example 3 – surface water supply with submersible pump

Description: an infiltration gallery next to the river is used to access the water source. The water from the infiltration gallery flows by gravity into a water tank which contains a submersible pump. This pumps the water to the house via a treatment system.

Controls: the pump is prevented from dry-running by a float switch in the water tank. A pressure switch switches the pump on and off according to water demand.

Example 4 – surface water supply using hydraulic ram

Description: the house is located above a fast flowing stream. A pipe in the stream supplies water to a hydraulic ram pump, which pumps a proportion of the water up to a storage tank which then gravity feeds to the house via a treatment system. The remainder of the water is returned to the stream.

Controls: the hydraulic ram operates continuously. An overflow pipe from the storage tank discharges into the stream.

In the examples with electrically powered pumps, pressure storage tanks may improve energy efficiency (see page 53), to prevent the pump turning on and off frequently.

Other components

Stop taps

Once you have your system installed, there will inevitably be occasions when you want to stop the water, so you will need to incorporate some valves (also known as appliance, isolation or control valves). You should try to have one upstream and downstream of each component of your collection and treatment system, so that any component can be isolated and removed. However, like all connections, valves add resistance to flow, and you may need to take this into account in your head loss calculations (as discussed in the box on calculating head loss). The internal design of the valve varies according to its function (and includes gate, cock, butterfly and ball) and in its resistance to flow. At the very least, you should make sure that there is a convenient tap near the house so you can turn off the water supply in an emergency.

Non-return valves

These valves allow water to flow in just one direction. They can be used as a backflow prevention measure under the Water Supply (Water Fittings) Regulations and are used in situations where there is a risk of one water source contaminating another (for example on outside taps).

Float valves

In a float valve, a float attached to a lever sits on top of the water. When the water level falls, the lever falls and opens a valve that allows water to flow. They are the type of valve used in header tanks and toilet cisterns, and are also used in a variety of storage tanks and water treatment systems. They are used when an air-gap is required to prevent contamination of a water source (and provide more protection against backflow than a non-return valve).

Foot valves

These are a specific type of non-return valve used at the bottom of supply pipes to suction pumps. They maintain a column of water in the water pipe that supplies the pump, so that it will not need priming prior to use.

Level sensors

If you want to know how much water is in your storage tank without looking in it, you can put a water level probe in the tank. These are available from electronic component suppliers. Circuit diagrams are available on the internet if you wish to make your own.

Air release valves

Air bubbles in the pipe will tend to join together and so may form air locks that can reduce flow or completely block a pipe. These are particularly likely to occur when a system is first commissioned, and will usually clear if sufficient water is run through the system; you might need a pump at this stage to get the water flowing properly, even in a gravity system. If the problem persists, you should check pipe joints (which can leak air into a pipe as well as leaking water out) or install an air release valve; both manual and automatic versions are available. Air release valves can easily become blocked open with suspended solids in the water, so must be checked regularly.

Water meters

If you are interested in working out how much water you use, you can buy a cheap inline water meter from a plumbers' merchant for around £20-£30.

Strainers

Vital to prevent solids damaging pumps, strainers are discussed on page 120.

Summary

Your water supply system will include some kind of inlet structure, storage, pipe-work, control mechanisms, cleaning processes and possibly an energy source. The most appropriate choice for each element is affected by a number of factors, including the type of source, how clean it is and the relative heights of the source and the house. These latter will need to be calculated, and sources of head loss accounted for. A suitably sized water store will need to be incorporated, and whilst open storage tanks are most common, pressure storage tanks can improve efficiency if an electrical pump is required. Renewably powered pumps are available, but their use is often site specific.

Further Reading

- *Water Distribution Manual* (Oxfam Water Supply Scheme for Emergencies). Considers the physics of pipeline design
- *Water Pumping Manual* (Oxfam Water Supply Scheme for Emergencies). Considers pump selection, installation and maintenance
- *Water Storage Manual* (Oxfam Water Supply Scheme for Emergencies). Oxfam guides are downloadable from the Oxfam website: www.oxfam.org.uk
- *Water Treatment & Sanitation*, Mann & Williamson, ITDG, 1982. Available from CAT Mail Order
- *Windpumps*, Barlow, 1993 – estimating the wind resource and specifying wind pumps for water supply. Available from CAT Mail Order
- *Introduction to Photovoltaic Power*, CAT Publcations, 2007. Available from CAT Mail Order
- *Tapping the Sun*, CAT Publcations, 2006. Available from CAT Mail Order
- Water Supply (Water Fittings) Regulations, 1999 – regulations that specify types of valves required according to contamination risk. Downloadable from

the HMSO website: www.hmso.gov.uk

• Water Regulations Guide, ISBN 0-9539708-0-9 – guidance from WRAS (Water Regulations Advisory Scheme) on the Water Supply (Water Fittings) Regulations, 1999

Chapter 4. Cleaning The Stuff

Deciding what water quality you want – getting it tested – what may be in it – removing solid contaminants – removing micro-organisms – removing dissolved materials – treatments for mains supplies.

Once you have found your water, you will need to decide how best to clean it. This is a very serious business; research into private water supplies in the UK has demonstrated that you are between 22 and 50 times more likely to contract a disease from a private water supply than from a public one, so you will need to ensure that you have the right type of system and that it is properly maintained. There are lots of methods you can use to clean water, and in order to choose the right one you will need to consider a number of issues.

What is the quality of the source water? How to clean water depends upon what is in it that needs removing. There are lots of different cleaning techniques designed to remove over 50 different contaminants. To complicate the matter further, water quality parameters can also be related to each other, for example, mildly acidic water isn't harmful in itself, but is much more likely to have metals dissolved in it than water that is neutral. Levels of nitrites and nitrates are also related to each other.

What water quality do you require? The basic assumption in this chapter is that you will want to treat at least some of your water to UK drinking water standards, and this quality is defined in law under the Private Water Supply Regulations (1991)[1]

Do you require all your water to be at this quality? Sometimes it makes sense to have separate water supplies of different quality. This might be the case if your source water needs a lot of cleaning, or you use large quantities of water for purposes that don't require high quality (e.g. garden watering, toilet flushing).

At what point between the water source and its end use are you going to clean it? You may clean the water close to the source and then store it, or you may store it first and then clean it immediately before use. As you will see, the choice is generally dependent on the type of treatment system you use, which in turn is dependent on the contaminants.

How can you do all of this in a sustainable way? Your primary concern will be to have a suitable quantity and quality of water, but the environmental impact of treatment processes varies considerably and you may wish to bear this in mind when choosing a treatment technique.

[1] In Northern Ireland, similar regulations were introduced in 1994. In Scotland, new (and significantly different) regulations were introduced in 2006 and are discussed in the Appendix

How many people would be affected by a failure in your water supply? For small scale domestic supplies serving single households, simple water treatment systems with only one or two elements, coupled with good source control are suitable, provided that they are maintained and operated correctly. However, if there are significant numbers of people using the water (e.g. a housing estate or a commercial enterprise), the principle of multiple barriers should be considered: that is, having a number of levels of control and treatment of the water supply, such that if a single treatment fails, water quality is not significantly compromised. This principle of multiple barriers should also include source control techniques (Chapter Two). Frequent monitoring of large systems is essential to ensure that problems are spotted and quickly rectified.

In this chapter, we will look at cleaning water from the perspective of the most likely contaminants and what treatment techniques we can use to remove them. The first reference point is therefore your initial water test results. If you are interested in treatment techniques per se rather than specific contaminants, Table 4.8 is provided for reference.

What do we want to achieve by cleaning the water?

What quality of water do you require?

The basic qualities of water that we are interested in are health, aesthetics (taste and colour) and corrosiveness.

Health

Water used for drinking and cooking must be safe to drink. This isn't simply about whether you have an upset stomach soon after drinking it; you could be slowly poisoning yourself with excessive metals, pesticides etc, without being immediately aware of the danger. So, just because the neighbours or the previous house owners say that they've drunk it for years without being ill, it may not be safe. It's not just drinking water that needs to be this clean; you may breathe in small quantities of water as a vapour when you wash, and your lungs can be a more vulnerable barrier than your gut lining – as evidenced by the risk of contracting *Legionella*, for example, from air conditioning systems or cooling towers where vapour is produced.

Aesthetics

Taste and colour are very subjective variables. You will probably expect the water that you drink to be completely clear, but you may be happy enough to flush your toilet with water that is slightly cloudy. In the same way, we have different attitudes and sensitivities to subtleties of taste, be they good or bad. Your attitude to both colour and taste may also vary according to whether or not you know what is causing it; if you knew that your water had a brown tinge because it originated from peaty soil you might look at it more favourably than water that was cloudy because it came from a grey water recycling system.

Corrosiveness

It is important that your water doesn't cause any physical damage to the pipes that it passes through, or anything that those pipes are plumbed in to, such as heating systems and sanitary appliances. This damage could be caused by the pH or the hardness of water, by blocking filters in appliances, or by abrasive action of particles in the water.

Different qualities for different purposes

As stated earlier, it may be appropriate to have separate plumbing systems within a building carrying water of different qualities for different purposes. This is especially true if you require large volumes of water of low quality for any purpose, or require a particularly intensive cleaning technique to make your water potable. But what quality of water do you need for what purpose? The Private Water Supply Regulations require water for drinking, washing, cooking or food production to be 'wholesome'. This is defined in chemical and microbiological detail, but basically requires that the water doesn't contain any substance at a concentration that is harmful to health, or any substance that may interact with another to cause it to be harmful. Water used for other purposes, such as toilet flushing, washing machines, heating systems and outside taps is not subject to the same legislation, and need not

Use	Regulation/guidance	Standard
Drinking, washing	Private Water Supply Regulations 1991	Various microbiological and chemical parameters (Table 4.4 and 4.5)
Toilet flushing	BSRIA proposed guidance	0 coliforms/100ml 90% of time <14 coliforms in any one sample
Swimming pool	WHO guidelines, British Standard CoP	<1 coliforms/100ml
Garden watering	Various, including WHO, WRAS, US agencies, German DIN19650	<1000 coliforms/100ml (WHO). <200 coliforms/100ml (WRAS) <2 coliforms/100ml (Several US states)

Table 4.1 Water quality for various purposes.

necessarily be treated to drinking water standard, but it should be clearly labelled to ensure that it isn't mistaken for a potable source. As an extreme example, much of Hong Kong is served by a dual water supply with one system carrying potable quality water, the other carrying sea water, which is used for toilet flushing, helping save precious potable water. Guidelines from a variety of agencies have been drawn up recommending water qualities for various uses, as shown in Table 4.1.

Getting your water tested

Since your choice of treatment technique will vary according to what's in the water, the first step will be to get a water sample tested. But how do you know what you should get it tested for? You will need to strike a balance; most tests are specific to an individual parameter so won't pick up things that aren't directly tested for, but when there are so many potential contaminants, testing for everything would be prohibitively expensive.

The requirements for private water supplies are set out in the Private Water Supplies Regulations 1991 (for Nothern Ireland in 1994 and for Scotland in 2006). These are based on the European Drinking Water Directive (80/779/EC), which in turn is based on guidelines from the World Health Organisation (WHO). The international nature of the guidelines means that many parameters that are listed are unlikely ever to cause problems in UK water supplies. The Private Water Supply Regulations are implemented by Environmental Health Officers in the local council, and specify both what to test for and how often. Both vary according to what the supply is used for and the number of people served by it, and supplies are divided into various classes, described in detail in the box on categories, below.

NB. These regulations are overdue an update. At the time of writing (early 2007), regulations governing supplies in Scotland have been updated, but not yet those for England and Wales or for Northern Ireland. The following discussion is based on regulations in England, Wales and Northern Ireland as they stand. The Appendix (page 167) discusses the new Scottish regulations. Updates to regulations for England, Wales and Northern Ireland are expected in 2007/8 and are likely to be virtually identical to the Scottish regulations. If in doubt about which regulations are in force, consult your Local Authority Environmental Health Department.

Under the 1991 regulations in England and Wales, small scale private water supplies are tested for 9 basic parameters, indicated in Table 4.4, but there are 43 additional parameters that may be tested for, should local factors (such as industrial or agricultural activity) make it necessary. The local officers will know from test results in the area what contaminants to expect and will organise the test for you (i.e. take the sample and send it to the laboratory). It is worth remembering that the test can only measure water quality at the time it is taken. For example, if you haven't got good source control measures (as described in Chapter Two), the quality may deteriorate after heavy rainfall that washes contaminants into your source. It is perfectly possible to take a water sample yourself and get it tested independently of local Environmental Health Officers, but if you do so, you should ensure that the laboratory doing the testing is accredited by UKAS (United Kingdom Accreditation Service), which issues NAMAS (National Accreditation of Measurement and Sampling) certificates to laboratories. Since the amount chargeable by the Council is limited by regulations, and they get preferential rates from labs, due to being regular customers, getting your own sample tested privately will be more expensive.

Categories of water supply

If your water is solely for private domestic use, it is a Category 1 supply. If the water is used as part of a business (e.g. camp site, holiday home, or small business), it is classed as a Category 2 supply. Within each category there are a number of different classes according to the volume of water used or the number of people served. For Category 1 supplies these are classes A to F, as shown in Table 4.2.

So, if you are a single household using water for domestic purposes only, then you will be a Category 1, Class F supply.

Category 2 supplies (where the private water supply is used for business purposes), are divided into Classes 1 to 5.

For more details on categories and classes, refer directly to the Regulations (which can be downloaded from the internet or purchased from HMSO publications), or contact your local authority Environmental Health Department.

Class	Number of persons supplied	Average volume of water (m^3/day)
A	> 5000	> 1000
B	501-5000	101 to 1000
C	101-500	21 to 100
D	25 to 100	5 to 20
E	< 25	< 5
F	One dwelling	

Table 4.2 Classes of water supply within Category 1 under the Private Water Supply Regulations.

Class	Average volume of water (m^3/day)
1	> 1000
2	101 to 1000
3	21 to 100
4	2 to 20
5	< 2

Table 4.3 Classes of water supply within Category 2.

Water supplies in Classes C, D, E, F, 3, 4 or 5 are tested for what the Regulations refer to as basic parameters and may also be tested for additional parameters depending on local conditions.

Basic parameter	Unit of measurement	Concentration or value (maximum unless otherwise stated)
Hydrogen ion	pH value	5.5 minimum, 9.5 maximum
Conductivity	µS/cm	1500 at 20°C
Total coliforms	number/100ml	0
Faecal coliforms	number/100ml	0
Lead	µg/l	50
Nitrate	mg/l	50
Odour	Qualitative	
Taste	Qualitative	
Turbidity	Formazin turbidity units	4

Table 4.4 Required standards of the basic parameters that will be tested for under the Private Water Supply Regulations. At the discretion of the local authorities (depending on local conditions) it may not be considered necessary to test for nitrate or lead.

Parameter	Concentration or value	Parameter	Concentration or value
Total Hardness	60 mg Ca/l (minimum)	Alkalinity	30 mg HCO_3^-/l (minimum)
Iron	200 µg/l	Sulphate	250 mg/l
Nickel	50 µg/l	Chloride	400 mg/l
Mercury	1 µg/l	Antimony	10 µg/l
Cyanide	50 µg/l	PAH	0.2 µg/l
Chromium	50 µg/l	Zinc	5000 µg/l
Cadmium	5 µg/l	Copper	3000 µg/l
Arsenic	50 µg/l	Trihalomethanes	various
Silver	10 µg/l	Tetrachloromethane	3 µg/l
Barium	1000 µg/l	Tetrachloroethane	10 µg/l
Fluoride	1500 µg/l	Colour	20 mg/l Pt/Co scale
Phosphorous	2200 µg/l	Aluminium	200 µg/l
Surfactants	200 µg/l	Nitrite	0.1 mg/l
Boron	2000 µg/l	Ammonium	0.5 mg/l
Total organic carbon	No significant increase	Odour (quantitative)	3 dilutions at 25°C
Oxidizability	5 mg O_2/l	Taste (quantitative)	3 dilutions at 25°C
Calcium	250 mg/l	Temperature	25°C
Magnesium	50 mg/l	Dry residues	1500
Sodium	150 mg/l	Colony counts	No significant increase
Potassium	12 mg/l	Manganese	50 µg/l

Table 4.5 Additional parameters under the Private Water Supply Regulations. These may be tested for if there is reason to believe that required levels may be exceeded. Values are maximum unless otherwise stated.

What do we mean by sustainable water treatment?

Water of any standard can be treated to drinking water quality, but it does not necessarily follow that it is appropriate to do so. Many environmentalists are interested in the idea of 'closing the loop', i.e. turning your wastes into resources. When applied to water, this could mean cleaning your sewage to make it safe enough to reuse. However, this is not necessarily an appropriate aim on a domestic scale. As discussed in Chapter Two, when you are assessing which potential water supply is best to use, the initial water quality is critical, as high quality source water will allow you to minimise the environmental costs of subsequent water treatment. Water treatment technologies have varying impacts on the environment, and since there is often a range of different methods you can use to clean your water, it can be difficult to choose the most sustainable technique. Things to watch out for include chemical use, energy requirements, waste water generated, components that need replacing regularly and infrastructure required.

Standards for water treatment techniques

As well as regulations on contaminants, there are a variety of certification schemes for water treatment units. Products bearing the label NSF have been certified to American standards, and in the UK some products are certified by WRc-NSF. BS EN standards also exist for many types of filter medium and treatment chemical, and the Drinking Water Inspectorate in the UK (DWI) has also produced definitive testing protocols for water treatment units. Treatment units bearing any of these labels will have been tested against a rigorous set of protocols, the nature of which varies according to the type of treatment unit (e.g. for UV units they involve passing water containing bacteria through the unit, whereas for cartridge filters they involve passing water containing specific sizes of dust through the unit). It is not a legal requirement to use water treatment products certified to these standards on your private water supply, but it is worth bearing in mind when assessing the merits of various systems, and units built in Europe will be manufactured to guaranteed standards. You should also be aware that most of the water treatment products that are advertised on the internet are designed to be used on mains water, not on private water supplies. If you read the small print it may well include a statement that they are not to be used with 'water that is microbiologically unsafe or of an unknown quality'. Consequently, they can be virtually useless for treating a private supply.

Categorising the contaminants in water

Once you have had your water test result, you can begin to consider how best to clean the water to the required standards. The following sections will consider how to control the various parameters, looking firstly at why the parameter should be controlled, and secondly what is the most effective way of controlling it.

At a basic level, water contaminants fall into three categories; solid things, living things and dissolved things, and this is generally the order in which a treatment system will be designed to remove them.

1) Solids

These are often as simple as bits of sand and soil. They may or may not do you any damage in and of themselves, but a soil particle can be home to millions of bacteria. Bits of sand and grit can also damage your plumbing system and appliances, impart colour to the water, and prevent certain types of treatment from working effectively.

2) Living organisms

These can cause serious illness, and can either be killed, or removed by filtration (since they are solids, albeit small ones, and they often live on the surface of larger solid particles).

3) Dissolved materials

There are lots of metals, minerals, gases and man-made contaminants that can be present in water. Some are dangerous to health in the long term, others impart flavour (pleasant or otherwise), and others affect more general properties of water such as pH and hardness, levels of which can be too low as well as too high. They can also affect colour, or damage appliances. Dissolved materials can be the most difficult type of contaminant to remove, and a variety of chemical and physical processes may be used.

Step 1 – removing solids

The properties of water which necessitate its treatment (health, aesthetics and corrosiveness) can all be affected by solids. Fortunately, solids are fairly easy to remove. The initial approach, as always, is source protection: that is, preventing solids getting into your water in the first place, as discussed in Chapter Two. For the purpose of water regulations, the amount of solids present in a water sample is indicated by turbidity. This is measured in formazin turbidity units: which is the amount of light scattered by a sample of the water compared to that scattered by a standard solution of a compound called formazin, and a limit of 4 units is prescribed. Solid removal can consist of one or several processes; settlement, flocculation and filtration.

Settlement

If a solid has a density very different to that of water, it will float or sink fairly quickly, but this isn't the case for many of the solids in water. Settlement times tend to be so long that it isn't worth constructing large settlement tanks or reservoirs specifically to settle out solids. The methods described in Chapter Two for getting your water out of the source should take advantage of any settlement that is occurring naturally.

Flocculation

Flocculation consists of adding a chemical to the water (usually aluminium sulphate) that encourages the solid particles to stick together to form larger heavier particles called flocs, which will sink more quickly. This process is also fairly effective at removing micro-organisms, as these are often attached to solid particles. Whilst it is routinely used by the water industry, flocculation is rarely appropriate for domestic scale water supplies owing to the careful control of chemical levels necessary and the complications of disposing of the metal-containing sludge which settles out.

Filtration

Filtering water is the most common technique used to remove solids on a domestic scale. A filter is basically a sieve, and the size of the holes in the sieve determines the size of particles that can pass through. The smaller the holes, the cleaner the water will be that comes out the other side. If the holes in a filter are very small, micro-organisms will be removed, since they are solids (albeit very small ones). However, smaller holes obstruct the flow more, so you may need to increase the pressure of the incoming water with a pump. The rate at which the filter will get blocked also varies according to the size of the holes; a fine filter will block more quickly than a coarse one. To avoid this, it is often best to have a coarse filter to remove larger particles, followed by a much finer filter. Some filter units have graded pore sizes to achieve this effect, with larger pores at the inlet than at the outlet. The speed at which water passes through the filter will also affect how many solid particles get filtered out; if you push water through a filter very fast, the solid removal will not be as good as if the water passes through more slowly.

The types of filters that are most frequently used on private water supplies are sand filters, carbon filters, ceramic filters, polypropylene/polyester cartridges and reverse osmosis units. Many filters are designed to fit under the kitchen sink, in a housing unit such as in Figure 4.2. These are straightforward to plumb in and are a standard size (10" or 20"), allowing a variety of types of filter to be used, depending on what contaminants you need to remove. They can also be plumbed in series (e.g. if several stages of treatment are required), or in parallel (if a high flow rate is required or to decrease the frequency at which cartridges require changing). As discussed in Chapter Two 'Finding a private water supply', several source collection techniques involve some preliminary filtration, so the choice of filter treatment will depend on the initial collection technique used (for example, infiltration used as a collection technique from a surface water supply will significantly improve water quality).

1) Sand filters

Figure 4.1 shows a slow sand filter. Water enters at the top, at a rate controlled by a ball valve, which is positioned to allow a 30 centimetres layer of standing water above the sand. The water passes down through a layer of filter sand 0.5-1.5 metres deep. Solids are removed by the simple sieving action of filtration, but an important additional characteristic of slow sand filters is that a population of bacteria and microscopic plants (known as a schmutzdecke), forms a fine mat in the top few centimetres of sand, and this mat of micro-organisms acts as a filter itself. Micro-organisms living in this layer will also feed on pathogens in the water, and UV from sunlight will also kill pathogens. Beneath the sand is gravel, which is separated from the sand by a layer of geotextile (to stop the two layers mixing). A slotted pipe in the bottom of the gravel takes the filtered water from the bottom of the tank, and this feeds into your water supply pipe. The rate at which water passes through the sand filter is controlled by a valve on the outlet pipe and should be a maximum of 100 litres/m^2/hour to provide good treatment. Special 'filter sand' should be used; grain sizes are in a narrower range (0.15-0.3 millimetres) and it is therefore less likely to block with 'fines' (the very small sand particles that are present in common sand). The sand filter should be covered to prevent debris falling into it, but since the action of UV light is beneficial, the cover should be some kind of mesh screen rather than a solid lid, unless there are lots of trees nearby (which might result in falling leaves, bird droppings and other sources of contamination).

Slow sand filters will remove most pathogens, but not reliably produce water with zero coliforms/100 millilitres (which is the requirement of the Private Water Supply Regulations). In situations where this pathogen removal is an important feature, a modified design of filter is recommended, in which the gravel layer at the base is only at the centre of the tank, with the outer 30 centimetres being sand. This means that any water that 'short circuits' the filter by flowing down its walls will be treated by sand in the base layer. The disadvantage of doing this is it can result in sand passing through into subsequent stages, which may damage pumps and

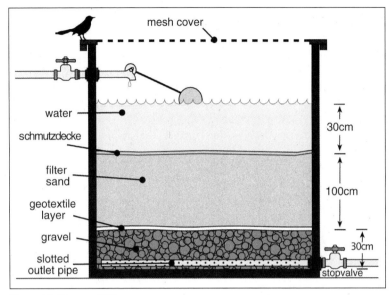

Figure 4.1 Slow sand filter.

appliances. Ensuring that the walls of the sand filter are rough rather than smooth will also help prevent short circuiting.

Two other types of sand filter are used in the water industry. Rapid gravity sand filters are used to remove floc particles (flocculation is described on page 81) and upflow sand filters are also used for solid removal, but neither are suited to domestic scale purification, since they require frequent backwashing and do not provide any biological action.

2) Steel coil filters

These consist of a coil of stainless steel (rather like the walking spring toy known as a 'slinky') through which the water passes. The surface of the coil has narrow ridges that keep each turn of the coil slightly separate, so that there are small gaps for water to pass through. Coil filters are available for various filter requirements, with 25 μm being the most appropriate for drinking water applications. They need backwashing at intervals, but automated units are available incorporating a differential pressure switch that senses when the pressure across the filter drops and then backwashes accordingly. If your water source is screened or settled, a steel coil filter is a straightforward, compact method of prefiltering water, before further treatment, to provide disinfection and ensure the removal of smaller solids. They are available to fit standard size cartridge housings.

3) Polypropylene, polyester filters

An older technology is that of the simple string filter. A variety of viscose, polypropylene, polyester and nylon filters is available, and filters may be woven, pleated or bonded (Figure 4.2(c)). As is the case for steel coil filters, string filters are best used as pre-filters prior to, for example, a ceramic candle. They are particularly appropriate for use with groundwater supplies where turbidity levels are already low. They are less suited to supplies with higher turbidity, since they can block quickly unless protected by a coarser pre-filter. They are not back washable, so must be disposed of when flow rate decreases. String filters are being superceded by steel coil filters.

4) Activated carbon filters

Activated carbon filters provide a physical filter medium that will remove solids (and can be specified for removal of particles >1 µm in diameter), but the advantage of carbon over other filter media is that it attaches contaminants to its surface by a physical process known as adsorption (which is not the same as absorption with a 'b'). They are therefore discussed in the section on removing dissolved substances (page 96).

5) Ceramic filters

Ceramic filters (also known as ceramic candles, owing to their shape) are an extremely good way of removing very small suspended solids including micro-organisms (Figure 4.3). Removal is quoted as >99.99% for particles of 0.9 µm, and >99.9% for particles 0.5-0.8 µm by a leading UK brand, and this can be assumed to provide a good level of treatment. This level of filtration provides excellent protection against micro-organisms, particularly in filters which are impregnated with silver (which kills micro-organisms trapped in the filter). Owing to the small pore size, a pre-filter may be required to prevent the ceramic filter blocking too quickly. The filter is housed in the same standard unit already mentioned (Figure 4.2) and can be removed from the housing for cleaning. Cleaning is required when the flow rate slows down; indicating that solids are blocking up the external pores. These solids can be removed with a clean scrubbing brush or scouring pad, making sure that you do not contaminate the inside of the cartridge with unfiltered water. The filter will need replacing every few years as you will gradually scrub off the surface of the ceramic. The case study on Undergrowth Housing Co-op in the box on page 95 indicates a typical installation.

6) Combined filters

Owing to concerns about micro-organisms breeding in filters and being re-released back into the water, many types are impregnated with silver. Lots of filters

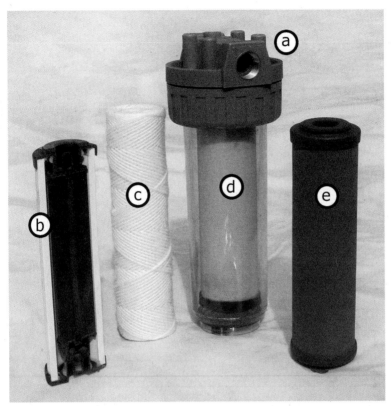

Figure 4.2 A range of filters available for (a) standard housing unit (b) cutaway version of a combined ceramic and carbon filter. The outer (white) region is ceramic and removes solids whilst the carbon core removes metals, pesticides etc by adsorption (c) string filter (d) ceramic filter (e) carbon filter (courtesy of Fairey Ceramics).

combine actions to remove solids, micro-organisms and dissolved substances. For example, ceramic filters can incorporate an ion exchange medium and a carbon core. This allows the filter to remove all solids including micro-organisms (ceramic and silver component), remove organic compounds and chlorine (carbon component) and to reduce heavy metal content (ion exchange resin). Similarly, carbon filters are available with various pore sizes and options that incorporate ion exchange resins or KDF (see page 96). Whether or not you need any of these additional functions will depend on the results of your water tests; it is pointless and wasteful of resources to have a filter that removes something that isn't there in the first place.

7) Reverse osmosis

Reverse osmosis (Figure 4.4) is a process in which water is pushed through a membrane with tiny holes in it (less than 0.001 μm). Its original application was to make sea water drinkable, and in countries with a scarce fresh water resource it is widely used on a municipal scale. Domestic scale systems are bad environmentally; you may need a pump to provide the necessary water pressure to push the water through the membrane and the waste water the unit generates can be 4 times the volume of the clean water it yields (the volume depends upon the initial salinity of the water). To prevent the membrane getting blocked, the unit includes prefilters (often both carbon and ceramic), so the water undergoes a good deal of treatment before it even reaches the reverse osmosis part of the unit. Water that has undergone reverse osmosis will be more acid than the source water (because the bicarbonate ions will have been removed), so may cause corrosion. Additionally, water purified by reverse osmosis is NOT recommended for drinking by the Department of Health (it is very soft, and there is some concern over a link between soft water and heart disease). Indeed, you can actually buy filters that 'remineralise' water that has been through reverse osmosis. The only situation in which reverse osmosis is appropriate for private water supplies is when there is no other way of removing excess dissolved materials (e.g. if you have water with a high salt content). Whilst it isn't possible for reverse osmosis units to be water efficient, you could consider reusing the waste water from the unit for non-potable purposes such as toilet flushing.

Filtration processes that rely on pushing water through a membrane are categorised according to the size of pores in the membrane. Nanofiltration and ultrafiltration use larger pore sizes than reverse osmosis, but are similarly unsuitable for private water supplies.

Step 2 – removing or killing the living organisms

Once we have removed the solids, the next stage is to remove or kill the micro-organisms. A micro-organism is simply a very small creature, and they are divided into a number of subgroups (bacteria, viruses, fungi, protozoa and helminths). Micro-organisms from all of these subgroups can cause disease, in which case they are called pathogens. Since there are lots of potential pathogens, it isn't practical to test for all of them, so instead we test for indicator organisms. The assumption is that any measure you take that kills indicator organisms will also kill any other pathogens that may be in the water. The indicator organisms used in the UK are total coliforms and faecal coliforms. Coliforms are a particular type of bacteria. Faecal coliforms are found in large numbers in the faeces of man and warm blooded

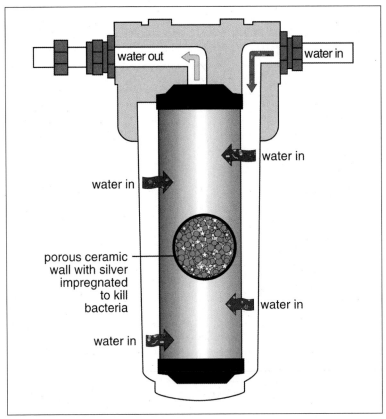

Figure 4.3 Ceramic filter. Water passes into the filter and solids are removed, including most micro-organisms. The filter is impregnated with silver which kills the micro-organisms that get trapped in it.

animals, so are a good indicator of faecal contamination; most of the pathogens that can be present in private water supplies come from contamination with sewage, either human or animal. Unfortunately, as indicated in the box 'Diseases from private water supplies in the UK' (page 88), some micro-organisms are more resistant to treatment than the indicator organisms.

There are two main approaches to removing micro-organisms from the water. The first is filtration; the micro-organism is a solid particle (albeit a very small one), so can be removed from water simply by passing the water through a fine filter. The second approach is to kill the micro-organism. This can be achieved with chemicals, ultraviolet light or heat treatment.

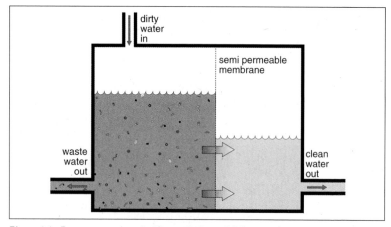

Figure 4.4. Reverse osmosis unit. If two solutions of different concentrations are separated by a membrane with tiny holes in it, then water will tend to move from the more dilute solution to the more concentrated one. If pressure is applied to the side with the more concentrated solution, the direction of flow will be reversed. This causes pure water to move through the membrane, but keeps contaminants on the input side. The water on the input side therefore becomes more concentrated and is discharged as waste.

Diseases from private water supplies in the UK

As stated in the introduction to this chapter, private water supplies are considerably more dangerous than the public supply in the UK. The most common symptom of disease related to water supply is diarrhoea and/or vomiting (often referred to as gastroenteritis), and in private water supplies this is most usually due to *Campylobacter*. These bacteria are easily removed by treatment and compliance with tests for the indicator organisms is an adequate safeguard. There are a number of other coliform bacteria that can cause gastroenteritis (or more serious disease), but source control and disinfection is a suitable safeguard against them.

There are few viruses that are problematic in UK private water supplies, with *Rotavirus, Norwalk-Like Viruses* (NLV) and *Eenteroviruses* being notable exceptions. They can be carried by birds and wild animals, and as always, the best protection is good source control to prevent contamination, and then adequate disinfection (e.g. by UV).

Unfortunately, some protozoan parasites are more resistant to treatment than the indicator organisms. The most relevant of these in the UK are *Cryptosporidium* and *Giardia*. Both are common in surface water, owing to the presence of animal faeces. Whilst most of us are fairly resistant to both parasites at low levels, both can be very dangerous to immune-suppressed individuals. Risks increase after heavy rainfall, since this causes agricultural slurry run off into surface water and hence increased levels of the protozoa in water supplies. Since they are not routinely tested for in private water supplies, preventing them getting into your supply in the first place is vastly preferable to trying to remove them once they are there. As with all micro-organisms, the best control measure is source protection as described in Chapter Two, although *Cryptosporidium* is effectively removed by fine filtration (e.g. in a ceramic filter).

It is worth reiterating that the major risk from private water supplies is from the huge number of supplies that still have no treatment; *Campylobacter* is a far more common problem than *Cryptosporidium* and is also easier to remove.

Filtration to remove micro-organisms

Most types of micro-organisms can be removed by filtration, as long as the pore size in the filter is small enough (most bacteria are > 0.8μm in diameter, so ceramic filters are ideal for micro-organism removal). Filters with larger holes will reduce levels of micro-organisms, since they are often attached to soil particles, although changes of flow rate (e.g. in sand filters) can wash some of these trapped micro-organisms through the filter). Many will also die during storage, as they are adapted to living in the human or animal body rather than in water. Ceramic filters are designed specifically to remove pathogens and will remove bacteria, fungi, protozoa (including *Cryptosporidium*) and helminths to safe levels (viruses will not be removed and are best controlled by source protection). Since micro-organisms that are trapped in the filter can breed and contaminate water passing through, most filters are impregnated with silver to provide a dual action of filtration and poisoning. In hard water areas, ceramic filters may suffer from premature blockage due to scale precipitation. Figure 4.5 shows the micro-organism removal profiles of various filter types; the only types that allow your water to consistently meet a zero coliform/100 millilitre limit are ceramic filters and reverse osmosis.

Killing micro-organisms

Killing micro-organisms is generally achieved by damaging their genetic material so they cannot reproduce, destroying them physically, or poisoning them with low concentrations of chemicals.

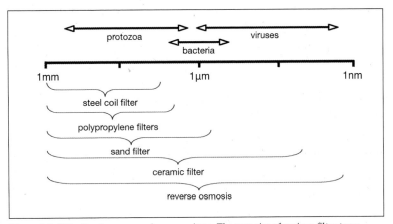

Figure 4.5 Filtration to remove micro-organisms. The pore size of various filter types affects their ability to remove micro-organisms. See text for the merits of different filter types.

Ultra violet light

A UV purification system for water is shown in Figure 4.6. It consists of a UV light (shaped like the fluorescent tube in a strip-light) in a quartz sleeve, which the water flows past. UV light damages genetic material, so that the micro-organism cannot reproduce. The sensitivity of micro-organisms to UV light varies according to the thickness of the cell wall; all common bacteria and viruses are effectively treated with the UV light intensities in commercially available treatment units, and recent evidence suggests that it is at least partially effective against Cryptosporidium and Giardia. Water must be clear for treatment to be effective (any particles will cast shadows that could shield micro-organisms from the UV source), and the flow rate through must be controlled to ensure adequate contact time. Build up of scale on the sleeve can also compromise effectiveness, so you should follow the manufacturer's guidelines on cleaning the sleeve. Unlike chemical poisons, UV does not provide any residual disinfection, so any micro-organisms that get into the water afterwards can multiply. In terms of sustainability, UV treatment is not ideal since it requires electricity. The bulbs need replacing regularly, since their brightness decreases over time, and can be difficult to dispose of since, like other fluorescent tube lights, they contain mercury vapour. Models are available with alarms to alert you to bulb failure. The UV unit should be put somewhere convenient as you will need to checck it regularly.

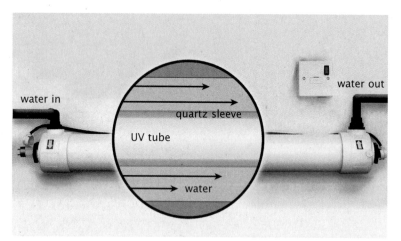

Figure 4.6 UV disinfection system. Water enters the sleeve around the UV bulb at one end and leaves at the other. The photo shows the bulb protruding slightly at the left end; the cutaway section shows what the inner sleeve and bulb would look like.

Chlorine, bromine, iodine

Devices that dose chlorine into water at a controlled rate can be used to disinfect water, but the environmental costs of chlorine production are high, and many people find the residual taste it leaves unpleasant. The main advantage of chlorination, which makes it the most widespread method for treatment of mains water supplies, is that it provides residual disinfection after treatment. Micro-organism levels will remain low during prolonged water storage and during transmission through the mains water network. However, it is poisonous to humans at high concentrations, and is complicated to use:

- it reacts with other contaminants in the water to form unpalatable (and potentially poisonous) substances known as disinfection by-products (DBP's)
- it is ineffective against Cryptosporidium
- the chemistry of chlorination is quite complex and varying doses are required to provide a suitable level of disinfection without taste impairment
- the presence of other contaminants and variations in pH will also impair chlorine's effectiveness

As a result, chlorination requires careful monitoring, and is rarely used for private water supplies in the UK. Infusion pumps that constantly drip chlorine into water are available, but rely on a continuous flow rate of water of constant quality. If you are planning to use chlorination for disinfection of your supply, you should obtain specialist advice from the equipment supplier. Other chemicals that are similar to chlorine are used on occasion; bromine is sometimes used in grey water systems, and iodine can be used for emergency water treatment, but neither is suited to regular use for treating water intended for drinking. Chloramines and chlorine dioxide may be used on large scale public supplies.

Disinfection of private water supply infrastructure

The components of a private water supply will need to be disinfected before use, to kill any pathogens that may be present, and it may also be necessary following maintenance. The final concentration in the structure you are disinfecting should be 0.2% chlorine by mass. Liquid bleach (sodium hypochlorite) contains about 5% free chlorine, so to make a 0.2% solution by diluting it you will need to add 1 litre of bleach to 24 litres of water. If the structure you are disinfecting already contains water (e.g. a borehole), you will need to work out how strong the solution should be to make a final concentration of 0.2% when you have added it to the existing volume in the borehole. If you can drain the structure (e.g. a spring box or a storage tank), it may be more appropriate to do this, and then scrub the walls with the solution, rather than attempt to fill the whole structure. The chlorine should remain in the structure for at least 12 hours, after which water should be run to waste until it no longer smells of chlorine. Suppliers of swimming pool chemicals can provide chlorine in tablet form (calcium hypochlorite), which may be more convenient.

Silver

As discussed on page 84, many filter units are impregnated with silver, so they have a dual action. Large micro-organisms are filtered out, and then killed by the silver, and smaller ones (including viruses) can pass through the filter but get killed by exposure to the silver as they pass through. The way in which the silver is bonded to the filter differs according to the type. In a ceramic filter, the silver is tightly bound and will not be released into the subsequent water to a significant extent; this is not the case with a carbon filter, where the silver is more easily released. However, this silver is not sufficient to provide any significant residual disinfection to the water. Silver is not appropriate as a domestic scale treatment technique other than in impregnated filters.

Ozone

The ozone molecule (O_3 as opposed to O_2, the type most common in air) is produced by subjecting O_2 to an electric field. Ozone is a very reactive molecule that will oxidise molecules on the cell membrane of micro-organisms, thereby killing them. Ozone can also remove organic contaminants from water supplies via oxidation. Ozone is increasingly used as an alternative to chlorination on municipal water supplies as it does not affect the taste of water. Just as with chlorine, the amount of ozone required varies according to the presence of other contaminants so working out a safe dose is complicated. Ozone is more effective than chlorine at removing Giardia and Cryptosporidium (though less so than filtration), but ozone provides no residual disinfection. Domestic ozonation units are available, but are expensive, have high power requirements and are not recommended.

Boiling and distilling water for removing micro-organisms

Boiling water for 1 minute will kill most micro-organisms and is a suitable treatment for water in an emergency situation. However, since it concentrates the water, levels of dissolved substances that can be harmful in the longer term will actually increase. Additionally, owing to the energy implications, it is not a sustainable solution for regular use. Distillation can also be used to purify water; steam from boiling water is collected and condensed, and passed through a carbon filter. This results in completely pure water, consisting of nothing but H_2O. Since minerals in water are important for taste, distilled water is not very palatable.

Step 3 – controlling dissolved substances – basic parameters

For many private water supplies, removing solids, followed by killing or removing of micro-organisms will be sufficient treatment. The other basic parameters that your water will be tested for under the Private Water Supply Regulations (listed in Table 4.4) are hydrogen ion content, conductivity, lead, nitrate, taste and odour. We shall therefore consider how to control these parameters.

Hydrogen ion content

Hydrogen ions (H^+) are responsible for acidity, which is measured on a logarithmic scale in pH units. A pH of 7 is said to be neutral. Lower than 7 is acid (lots of H^+ ions), higher is alkali (fewer H^+ ions), and the limits in the Private Water Supply Regulations are a minimum of 5.5 and a maximum of 9.5. The pH level itself is not bad for you, but affects what may dissolve in the water, and both high and low pH can be extremely corrosive to your plumbing system. Water with a low pH is likely to dissolve metals such as iron and manganese, and this is a common problem with upland surface water sources, which will not have passed through carbonate-containing rocks that would neutralise the acidity. Sometimes organic acids from the breakdown of vegetation can also cause source water to be acid. The pH can be raised (made more neutral) by passing the water through a filter that contains dolomite (limestone), which dissolves into the water and imparts alkalinity. The dolomite filter can be an under-sink unit, or in a separate filter rather like a slow sand filter (as used in the Undergrowth Housing Co-op case study on page 95). Loose dolomite chippings in your storage tank will also help neutralise the water. Other compounds (e.g. magnesium oxide) can also be used, although some systems are prone to sedimentation and are not appropriate for irregular use (e.g. in holiday homes), so you should seek advice from the manufacturer.

Conductivity

As you can see from the table of additional parameters (Table 4.5), there are lots of things that could be dissolved in water. Most chemical tests are specific for individual chemicals, so you would need to perform lots of different tests on a water sample if you wanted to know exactly what was in it. Luckily, a fairly non-specific test is available to give an initial idea of whether there is anything in the water that is of concern. Conductivity (measured in micro Siemens/cm, with a limit in the UK of 1500) is a measure of the total concentrations of positively and negatively charged ions in the water and since lots of contaminants can exist as ions, conductivity is a good general test for the presence of contamination. Pure H_2O has low conductivity, whereas water with lots of ions in has high conductivity. Most ions likely to be present (such as sodium, potassium and chloride) are perfectly safe at levels that cause high conductivity, but more harmful ions such as arsenic, lead and cadmium could also be responsible for a high conductivity reading. Consequently if your water has a high reading, you will need to have tests carried out on the additional parameters (detailed in Table 4.5) to find out what is causing it. A high conductivity reading could also be an indicator of saline intrusion into your groundwater supply, in which case you may well have to find an alternative water source.

Lead

The limit for lead in private supplies is 50µg/litre, based on the risk of brain damage for higher levels of consumption. This is higher than the limit for public water supplies, and will be reduced to 10µg/litre in forthcoming regulations on private water supplies. Lead can be from your plumbing or due to local geology. Lead was phased out for domestic plumbing around 1970, so if your plumbing is more recent, then it should not contain lead. You can tell if you have lead pipes by scratching slightly at the surface of the pipe where it comes in to the house: lead is a dull grey that becomes shiny when you scratch the surface off. If you find lead pipes, remove them. If your water contains lead due to the local geology (a local history of lead mining is an obvious indicator!), you should check the pH of the water; acidic soft water will dissolve lead, so you may be able to reduce lead levels sufficiently simply by neutralising the water. Ion exchange media can also remove lead.

Nitrate and nitrite

Nitrates from fertilisers have found their way into our surface water and groundwater in many areas following years of intensive agriculture. Nitrate is also a breakdown product of ammonium, so can occur in waters that have been contaminated with sewage. High levels of nitrate are associated with 'blue baby syndrome' (methaemoglobinaemia) in bottle fed infants, and early-onset diabetes. The limit in drinking water is 50 milligrams/litre, although a reading of more than 25 milligrams/litre is often regarded as requiring treatment. Nitrite can cause the same problems, but at much lower concentrations (the limit is 0.1 milligrams/litre). The simplest way to remove nitrate and nitrite on a domestic scale is with a nitrate specific anion exchange resin; water is passed through a unit that swaps the nitrate ions for chloride. These units are available as standard sized cartridges.

Taste and odour

Qualitative assessments of taste and odour can tell you a lot about water quality and can indicate whether more specific tests need to be carried out. Solid organic material (algae, moulds, bacteria and breakdown products from plant material) can result in an unpleasant taste, but will have largely been removed from the water by filtration, but there are also some dissolved organic breakdown products that can cause taste problems. These are generally removed with a carbon filter. Organic material can also result in odour.

Many other substances that are listed as additional parameters have limits based on their effect on water acceptability rather than due to any health effects, and will be considered below.

Case Study – Undergrowth Housing Co-op

Undergrowth Housing Co-op is in a remote location in mid-Wales. It is 220 metres above sea level, and mains for both electricity and water are more than two miles away. A stream runs past the house, and a small reservoir has been built that is used to run a hydro electric turbine. The reservoir is 11 metres above the level of the house; this is not enough to provide sufficient water pressure for the household water supply once a cleaning stage has been included, unless a pump is installed. Instead, water is taken from the stream that feeds the reservoir at a point 18 metres above the level of the house. When analysed, this water had a low pH and consequently high levels of several metals (see Table 4.6), in addition to high levels of coliforms and solids (as expected for upland surface watercourses).

The immediate action was to install a ceramic filter underneath the kitchen sink (to remove the solids and the coliforms). When the residents had more time, a slow sand filter and storage tanks were constructed 15 metres above the house, to provide solids removal and decrease the frequency at which the ceramic filter became blocked. To neutralise the water (and so reduce the metal levels), an additional filter tank containing dolomite chippings was installed. Since installation, the water quality has improved (see Table 4.6) and all metals levels are well within limits.

Contaminant	Permitted levels	Before treatment	After treatment
pH	5.5-9.5	6.04	7.2
Manganese	50µg/l	116	7
Aluminium	200µg/l	290	65
Iron	200µg/l	469	63

Table 4.6 Effects of water treatment at Undergrowth Housing Co-op.

Step 4 – additional parameters under the Regulations

The Private Water Supply Regulations include a list of 43 additional parameters, which may be tested for if there is reason to believe that levels may not fall within prescribed limits. Most of these are dissolved substances and are listed in Table 4.5. Many have their basis in international guidelines and are rarely a problem in UK water supplies. It is also worth making a distinction between those that affect health and those that adversely affect taste at levels well below that which can affect health.

Ion exchange resins to control additional parameters

Many of the additional parameters can be controlled using a technique called ion exchange. A variety of units is available, depending on the parameter to be controlled, but the principles behind them are the same. The ions (charged elements or compounds) that are impurities in the water are replaced by ions from the ion exchange resin. There are two basic types of ion exchange resin; cation exchange resins which exchange positively charged ions in the water for H^+ ions in the resin, and anion exchange resins which exchange negatively charged ions in the water for OH^- (or sometimes other anions) in the resin. If you need an ion exchange resin, you should make sure that the one you purchase is capable of controlling the parameters you are concerned with; ion exchange resins don't all do the same thing.

Activated carbon filters

Activated carbon filters trap contaminants in the filter by a physical process known as adsorption (which is not the same as absorption with a 'b'). This means that dissolved substances that would otherwise pass through the filter will bind to the carbon and so be removed from the water. They are called 'activated' carbon filters because of the way in which the filter medium is produced, resulting in a very high surface area per gram of material ($500\text{-}1500\text{m}^2$/gram) (rather like that in a honeycomb), providing a lot of binding sites for material to be adsorbed on to. There are two main types of activated carbon filter; granular activated carbon (GAC) and block carbon, and they are most often used in private supplies to remove pesticides and herbicides. They are also effective at removing chlorine and improving taste, so they are often used in jug filters designed for use with mains water supplies. Block carbon filters have a more compressed structure, but require higher water pressures and prefiltration to prevent premature blockage. Both types must be regularly replaced (frequency varies according to the manufacturer): once the binding sites on the surface of the carbon molecules are used up the filter can actually start releasing the contaminants back into the water again. It isn't practical to clean carbon filters; they must be thrown away after use. The most common types of carbon filter are available to fit a standard filter housing that is positioned under your sink (as illustrated in Figure 4.2), and they often incorporate a small filter on their outlet to trap fines of carbon that may be released from the cartridge.

KDF filters

A compound increasingly used in combined filters is 'KDF-55D', a copper-zinc alloy that relies on redox reactions (reduction-oxidation, a chemical term that need not concern us here). This removes a range of metals including lead, iron, mercury, nickel and chromium. It also kills bacteria (although not to zero coliforms/100 millilitres). Often used in combination with granular activated carbon and an ion exchange resin to produce an extremely wide spectrum of contaminant removal.

Naturally occurring substances affecting health

Arsenic, boron, barium, chromium

The first indication that you may have a problem with these metals is likely to be a high conductivity result. Arsenic contamination can be natural (e.g. as has been discovered in Bangladesh), or as a result of mining, or in areas where surface water run off is contaminated with arsenic used as an industrial poison. Boron and arsenic can be removed by activated carbon filters, and all four of the above elements can be removed by ion exchange or KDF filters (discussed on page 96). Limits for these substances are given in Table 4.5.

Radiation

Naturally occuring radon and uranium are found in groundwater in certain parts of the UK (e.g. the south west of England), depending on the underlying geology. Radon is sufficiently volatile that aerating the water (in a well ventilated area) can reduce levels sufficiently. It can also be removed by carbon filters. Uranium can be removed by ion-exchange, although finding an alternative water source is preferable.

Naturally occurring substances affecting acceptability

The following limits are based on their effect on aesthetics rather than on health. However, those substances that cause cloudiness (e.g. aluminium) or staining (e.g. iron, manganese) may interfere with treatment processes such as UV disinfection that rely on the water source being clear.

Water hardness

As rain falls, it dissolves carbon dioxide from the atmosphere and becomes a weak solution of carbonic acid. This slight acidity in the water means that it will dissolve minerals such as calcium and magnesium in the rocks that it passes through when it hits the ground. Groundwater supplies in areas with calcium or magnesium-rich geology, and surface water supplies fed by groundwater in these areas, are said to suffer from 'hard' water. Areas of the UK with hard water are illustrated in Figure 4.7. The calcium and magnesium carbonates that cause hardness can precipitate out as scale in pipes and appliances. They increase the amount of soap required to produce lather, and cause detergents to create scum. Consequently many people are interested in softening their water, and this is discussed on page 105. However, it is worth bearing in mind that regulations on hardness actually specify a minimum allowed hardness, with no maximum value; hardness is NOT a bad thing from the perspective of drinking water, as very soft water can be very corrosive to plumbing systems and so may release harmful chemicals into the water from pipes and

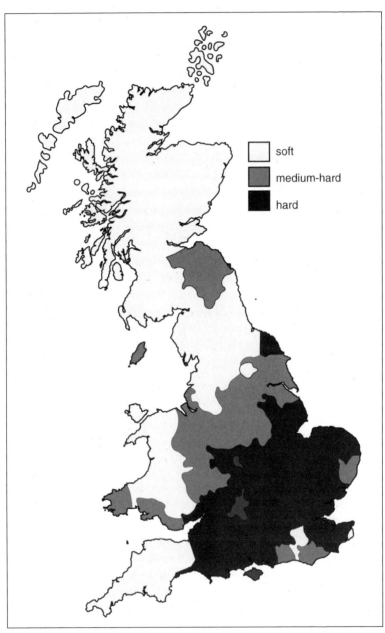

Figure 4.7 Hard water areas (courtesy of Avonsoft).

fittings. There are, however, maximum levels for both calcium and magnesium.

Iron and manganese

The limits on iron and manganese in the Regulations are based on amounts that impair taste or cause staining, and are far below those that cause health problems. Both metals can exist in two forms; one soluble, the other insoluble, and the most common removal techniques are designed to oxidise the metal so that it changes into its insoluble form. The pH of water dramatically affects iron and manganese content; if your water is acidic, simply neutralising it will cause iron (and to a lesser extent manganese) to turn into its insoluble form, after which it can be removed by filtration (as demonstrated in the case study of Undergrowth Housing Co-op on page 95). Alternative approaches involve oxidising the metals either by aeration or by chlorination, or by adsorption onto manganese dioxide. The solid manganese can then be removed by filtration. KDF filters (on page 96) will also remove iron and manganese.

Calcium, magnesium, sodium, potassium, sulphate, chloride

These ions affect taste at levels well below those affecting health and are rarely a problem in UK water supplies, unless there is saline intrusion into a groundwater source. They are often tested for if conductivity levels are high (as described on page 93), along with other more dangerous ions. They can be removed by ion exchange if necessary, although if saline intrusion is suspected and the water is brackish, then distillation or reverse osmosis are the only options.

Ammonium

The Regulations set an ammonium standard of 0.5 milligrams/litre. Ammonium contamination of a private water supply is generally due to the presence of sewage (human or animal), so you should double check your coliform test results if your water fails on ammonium. If your water tests fail for both coliforms and ammonium, you should control the contamination at source, rather than trying to clean the water after it has become dirty. Ammonium is removed by oxidation (which can be achieved using aeration). Since the oxidation of ammonium produces nitrates, you may well need to control nitrate levels in this instance. On a large scale, ammonium is removed using chlorine, but this results in the formation of chloramines which in turn can cause taste and health problems.

Hydrogen sulphide

Hydrogen sulphide produces a characteristic 'rotten egg' smell and is common in water with high sulphur content. It can be removed by oxidation (aeration), with an under-sink KDF unit, or by addition of chlorine.

Chemicals from human activity affecting health

Antimony, zinc, copper, cadmium, mercury and nickel

These are rarely found at elevated levels in drinking water supplies unless there has been substantial mining or metal working in the area, or if they are from water pipes and fittings. They can generally be removed by filtration through activated carbon or ion exchange systems, but check with the manufacturer of the filter that what you are buying is designed to remove these elements.

Fluoride

Few waters contain high levels of fluoride in the UK (either naturally or due to industrial pollution), although the eastern side of the UK has higher natural fluoride levels compared to central and western areas. Of more general concern is the fact that a number of water companies add fluoride to public water supplies, and this is discussed on page 104. If your private water supply does contain high levels of fluoride then it can be removed using aluminium oxide cartridges, available for standard filter housings (fluoride is also removed by reverse osmosis, but this has a higher environmental impact).

Trihalomethanes, tetrachloromethane, tetrachlorethane, PAHs

A variety of carcinogenic man-made organic materials can be found at low levels in water. Some of these compounds are organic solvents that enter the water system by accidental spillage; others are produced during water treatment with chlorine, in which case they are known as disinfection by-products, (DBPs). The other class of organic compounds of particular concern are the polynuclear aromatic hydrocarbons (PAHs). These are formed during combustion and can enter drinking water through old mains water pipes lined with coal-tar (although a more likely source of high exposure is cigarette smoke or roasted, fried, grilled or baked food). You are unlikely to have high levels of any of these compounds in your private water supply. However, if you do find them in your water, they can be removed by adsorption in activated carbon filters.

Aluminium

Aluminium is used as part of the treatment process in public water supplies, but is not found naturally in water at significant levels. Links between aluminium and Alzheimer's disease have been researched for several decades, but even the Alzheimer's Association agrees that no causal link has yet been demonstrated. Since aluminium from drinking water will normally comprise less than 5% of total dietary intake, there are easier ways of reducing your aluminium intake, if you are at all concerned about it, than by trying to remove it from your drinking water. Aluminium is present in tea bags, anti-perspirants and at high levels in many antacid tablets.

If you wish to remove it from drinking water, specialist cation exchange resins are available. Before resorting to this, check the pH of the water. If it's acid, neutralising the water (page 93) may well reduce aluminium levels sufficiently .

Maintenance of private water supply treatment systems

As discussed earlier in the chapter, a good water test result is no indicator of future water quality. Since many treatment techniques (e.g. ion exchange) will tend to decrease in effectiveness over time, and all will vary according to the quality of the source water, it is worth considering how frequently elements of the system require maintenance or replacement. Generally, this information will be supplied by the manufacturer, but Table 4.7 is a good starting point for the most commonly used technologies.

Other elements of the water supply system will also need checking regularly. Inlet devices should be checked to see that they are not blocking (surface water supplies) and that source protection measures such as fences, lids etc (groundwater supplies) are in place (as discussed in Chapter Two). Storage and header tanks should also be inspected periodically to make sure that water quality doesn't deteriorate during storage.

If you already have a private water supply

If you already have a private water supply, you should have it tested periodically to see what's in it. The local authority may do this anyway, but if you have a Class F supply (single domestic household, with solely private water use) they are not obliged to ever test your water (although they have the right to do so if they wish). A history of safe use can indicate that pathogen contamination is low, but it could equally well mean that the people drinking the water have built up immunity to the pathogens that are present, or are happy to attribute infections and illnesses to other causes. Remember, you are up to 50 times more likely to contract a disease from a private water supply than a public one! Since a water quality test is just a measure of quality at that time, it is no guarantee of future quality, or quality under different conditions, e.g. after heavy rainfall (which frequently leads to higher numbers of micro-organisms in supplies). It is also no guarantee of chemical safety; you may be slowly poisoning yourself with lead, arsenic, nitrates, pesticides etc. The best way of assuring water quality is to practice the source control measures detailed in Chapter Two, and to refer to the manufacturer's guidelines (and Table 4.7) on maintenance frequencies required for various technologies. We would recommend that you get your water tested every year, whenever collection or storage circumstances change,

Treatment technology	Maintenance requirement
Slow sand filter	Remove top 3cm of sand every 4 months or when flow rate decreases
Ceramic filter	Scrub clean (using aseptic technique) when flow rate decreases. This varies according to the turbidity of the water, but is often required every 3 months. Replace when the cartridge shows visible signs of being eroded by the cleaning process
Carbon filter	Replace at intervals recommended by manufacturer (often 6 months)
Steel coil filter	Backwash weekly, remove from housing and clean annually
Polypropylene, polyester, viscose, nylon filters	Replace when flow rate decreases, or at intervals recommended by the manufacturer (often 6 months)
Ultraviolet	Replace bulbs at intervals recommended by manufacturer (often annual)
Ion exchange resins: softeners	Replenish salt solution at intervals recommended by the manufacturer
Dolomite filter	Refill with media at intervals recommended by the manufacturer (under-sink versions), or when quantity of dolomite chippings has halved (DIY versions). Alternatively monitor pH and add dolomite accordingly
Ion exchange resins	Replace at intervals recommended by the manufacturer (often 6 months)
Copper-zinc alloy (KDF)	Replace at intervals recommended by the manufacturer (often 6 months)

Table 4.7 Maintenance requirements for a range of treatment systems.

or if you notice any unusual discoloration, taste or odour.

When to clean the water

As well as how to treat the water, you will need to consider when to treat it. At its simplest, this comes down to whether you treat water before storing it, or afterwards. Treatment techniques that operate slowly (such as sand filters) will need to incorporate storage buffers to help match supply and demand. However, you should be aware that there is the potential for water to become contaminated during storage, so there is much to be said for treating the water immediately before it is used. Systems treating dirtier water may well have a system providing basic cleaning after which the water is stored, before passing through another system (e.g. an under-sink unit) within the house (for example as in the case study on Undergrowth Housing Co-op, page 95). In practice, your choice will normally be governed by what treatment technique you are using (and how many stages), which in turn will depend on the contaminants. Another major factor is whether the raw untreated water (or water that has gone through just one stage of treatment) is of good enough quality to use for washing and bathing. If this is the case, then simply treating water to drinking standards at the kitchen tap will be sufficient, as long as everyone is aware that they should not use water elsewhere in the house for drinking. For example, it may be appropriate to have sand filtered water serving most of the house, with the additional protection of a silver impregnated ceramic filter for the kitchen tap (as in the case study of Undergrowth Housing Co-op).

Treatments for mains water supplies

Filtering mains water and drinking bottled water are becoming increasingly common. If you aren't happy drinking your tap water, you should start by deciding whether this is for reasons of health or of taste.

Taste

The taste of water in the public supply varies considerably across the UK, and it may also vary according to the time of year, as water companies switch from one source to another. Taste is a very subjective measure, but is often significantly improved by leaving the water in the fridge for a few hours; it may taste less 'flat' and any chlorine taste is less noticeable, so this should be your first resort. As discussed below, storing water in the fridge also decreases levels of disinfection by-products. If leaving it in the fridge doesn't do the trick, filters (either in a jug or plumbed in under the kitchen sink) can be used. These are usually carbon filters and their main taste-altering action is the removal of chlorine. However, when you remove the chlorine, you also remove its residual disinfection effect. Bacteria can actually breed in water filters and poorly maintained filters may increase bacteria above levels in the incoming tap water. The adverse taste effects of chlorine are causing some water suppliers to move to use of chloramines and/or ozone.

Health

The public water supply in the UK was sampled almost 3 million times in 2005 and 99.96% of those samples passed internationally approved tests. In contrast, routine monitoring of private water supplies suggests that 50% regularly fail similar tests. There have also been high profile incidents of bottled water being found to be contaminated; indeed bottled water can pose a particular risk since it rarely contains any residual disinfection. Nevertheless, some people are concerned about the effect on health of a few key parameters in mains water supplies.

Fluoride

A number of water companies add fluoride to the public water supply at the request of the local health authority (artificially fluoridated water supplies about 10% of the UK population). Overwhelming evidence indicates that low levels of fluoride offer protection against tooth decay, but opponents of fluoridation would argue that most people get enough fluoride from their toothpaste and food to achieve these beneficial effects. Studies have failed to find conclusive links between fluoride intake at normal levels and adverse health effects, although prolonged, artificially high intake has led to adverse effects on teeth and bone. If you are worried about fluoride, you should start by removing other sources of fluoride from your diet (fluoride levels are high in seafood, toothpaste, tea and soft drinks). If you wish to remove it from your water supply as well, you can install an activated alumina filter cartridge that fits in a standard housing unit (Figure 4.2). Regardless of the health effects of fluoride, the ethics of what is in effect widespread semi-compulsory medication may be of some concern.

Chlorine and disinfection by-products

One of the drawbacks of using chlorine for treatment of mains water supplies is that it reacts with organic substances in the water, forming a range of compounds termed disinfection by-products (DBPs), some of which are harmful (e.g. chloroform and chloroacetic acids) at high concentrations. Public concern surrounding the issue often relates more to the fact that you can taste them than to any health effects; the concentrations of DBPs that adversely affect health are far higher than those at which adverse taste is experienced. However, if you are concerned, you can ask the water company to do a test of your water, and you can also ask them for results of their sampling, so you can compare them to the levels indicated in Tables 4.4 and 4.5. Storing water in the fridge significantly decreases concentrations of several DBPs, as well as improving taste. If you are still not happy, consider installing an activated carbon filter (as described on page 96) to your kitchen tap. Some public water supplies now use chloramines (as opposed to chlorine) as this provides more robust disinfection and produces a less noticeable taste. Unfortunately, they are also considerably more difficult to remove using domestic water filters.

Lead

The limit for lead in public drinking water is 25µg/litre, falling to 10µg/litre in 2013. You should check to see if you have lead pipes in your house (as discussed on page 94), and your local water company will be able to tell you how much lead is in the mains water, which by law is closely monitored and controlled. In areas known to have lots of lead pipe work, phosphate may be dosed into the water supply. This stabilises the lead pipe by forming a layer of lead phosphate scale on the inside, which prevents lead dissolving into the water. The pH of water is also closely controlled to minimise the amount of lead dissolving.

Water softeners

As discussed on page 97, water hardness is caused by calcium and magnesium carbonates dissolved in water. These compounds can precipitate out as scale in pipes and appliances. They increase the amount of soap required to produce lather, and cause detergents to create scum. Consequently many people are interested in softening their water. However, artificially softened water is potentially bad for your health, owing to the high levels of sodium it contains; if you install a water softener on your mains supply, you must also have a dedicated drinking water supply that has not been softened, as there is a link between high sodium diets and cardiovascular disease. Additionally, there may be a protective effect against cardiovascular disease from the higher levels of calcium and magnesium in hard water. Softened water should never be given to bottle fed infants, or people on low sodium diets. Soft water can also be corrosive to your plumbing.

The concern with hard water, from an environmental perspective, is the fact that a 1mm layer of scale can reduce the efficiency of heat transfer by around 20% and so can seriously impair the efficiency of an immersion heater element. If you live in a hard water area, you should remove the element and de-scale it periodically in preference to softening your water. Advice on water softening for laundry use is more equivocal; there is limited evidence from Life Cycle Analysis (LCA) suggesting that the environmental costs of water softeners can be justified by the decreased use of laundry detergents softer water allows.

There are two basic approaches to dealing with water hardness. Water softening units are based on cation exchange; water is passed over a resin in which the calcium and magnesium ions that cause scale are replaced with sodium ions from the resin. The unit includes a tank containing salt water that automatically backwashes the resin periodically, to recharge it with sodium ions. Other types of unit (including magnetic, electrolytic and electronic) are known as 'scale inhibitors' or 'physical conditioners' and do not actually soften water, but a few studies have suggested that they reduce the likelihood of scale forming. Results with these units are variable; you should ensure you are offered a money back guarantee if you are purchasing a scale inhibitor.

Treatment	Comments
Removal of solids	
Settlement	Useful initial stage, but large storage volumes necessary.
Flocculation	Difficult to control; by-products difficult to dispose of; not suited to domestic scale.
Sand filtration	Effective and simple removal of solids and most micro-organisms.
Ceramic filter	Effective and simple removal of solids and micro-organisms, but may require preliminary coarse filtration to avoid filter blocking prematurely.
Carbon filter	Not effective as sole method of solid removal, more commonly specified for removal of dissolved materials such as pesticides and solvents. Often used to improve taste and odour on mains water supplies.
Steel coil filter	Compact (under-sink unit) and suitable for use as a pre-filter.
Polypropylene, polyester, viscose, nylon filters	Available as woven, pleated or bonded cartridges. Useful as a pre-filter (e.g. to a ceramic filter), but not suited to water with high turbidity (e.g. surface water), when an additional pre-filter would be required.
Removal of dissolved materials	
Carbon filter	Removes halogenated organic compounds and chlorine. Often combined with ceramic filters for good all round treatment.
Ion exchange resins: softeners	Remove calcium and magnesium and replace with sodium. Resultant water is good for washing, but not for potable purposes due to high sodium.
Dolomite filters	Neutralise acidity; may facilitate iron and manganese removal.
Ion exchange resins: other cations	Removal of heavy metals, such as lead, arsenic, barium.
Ion exchange resins: anions	Remove nitrate and replace it with chloride. Will also remove sulphate,but if sulphate levels are high, nitrate removal may be inadequate.
Copper-zinc alloy (KDF)	Removes iron, manganese and heavy metals. Some micro-organism removal (but insufficient as sole measure).
Reverse osmosis (RO)	Expensive, wasteful of water, energy intensive. Pre-filters in RO units (ceramic and carbon) often result in good quality water with little additional benefit derived from the RO. Not suited to drinking in the long-term and often unpalatable.

Table 4.8 (above and opposite) Techniques for cleaning water and the contaminants that they are designed to remove

Treatment	Comments
Removal of living organisms	
Chlorine	Good disinfectant, but can react with organic compounds to produce carcinogenic disinfection by-products. Difficult to control dose if levels of organic contaminants vary. Imparts undesirable taste, but provides residual disinfection, so allows storage.
Bromine	Not used for potable water owing to its toxicity. Sometimes used in grey water systems and swimming pool disinfection.
Iodine	Effective for emergency disinfection, but not recommended for long-term use due to its adverse health effects.
Ultraviolet	Effective disinfection, via disrupting the DNA of micro-organisms. Water must be clear for UV to be effective. No residual disinfection. Bulbs contain mercury, complicating end of life disposal.
Ozone	Dangerous by-products formed during treatment, not suited to small scale use.
Ceramic candle with silver	Ceramic provides filtration, with trapped micro-organisms killed by silver. No residual disinfection.
Boiling	Suitable for short-term use only. Energy intensive and often makes water unpalatable. Concentrates dissolved ions.
Distillation	Suitable for short-term use only. Energy intensive and often makes water unpalatable. Removes minerals from water.

Bottled water

'Nutritional information' labels on bottled water consist largely of data that is fairly irrelevant to health; the contents must comply with the same parameters that public and private supplies do. There is no health benefit to be derived from drinking mineral water: the main difference between bottled water and tap water is price; bottled water can easily be 3000 times more expensive and have been shipped round the globe. Indeed, since bottled water does not contain residual disinfectant, and is stored for longer periods of time and at higher temperatures than water in mains distribution systems, you could argue that it is actually more likely to contain pathogens. There have been a number of high profile cases of bottled water being contaminated by both pathogens and dissolved substances. Because of this lack of residual disinfectant, bottled waters are routinely tested for more microbiological parameters than the water in the mains supply. The higher mineral content of some bottled waters makes them unsuited to infants and young children, however, manufacturers are not required to state this on the label. Clearly, if you need a bottle of water, the best thing to do is to fill one from your tap.

Reference table on treatment techniques

As discussed in the introduction to this chapter, the technique you use to clean water depends upon the parameters you are trying to remove, but if you are interested in techniques per se, Table 4.8 provides a basic guide to what can be removed by each one.

Summary

The way in which we clean water depends on the parameters we are trying to control, and these in turn will vary according to the end use of the water. The first stage, therefore, is always to get a sample of the water tested. The contaminants of concern are usually solids, pathogens and dissolved substances, and water is generally treated for each of these in turn. Filtration is often the most suitable form of treatment for domestic supplies, and a range of filter types with various properties is available to remove a variety of contaminants.

Further Reading

- *WHO Guidelines for drinking-water quality*, Third Edition, 2004, ISBN 92 4 154638 7 - the worldwide definitive (500 page) document on water quality. Downloadable from WHO website: www.who.int/water_sanitation_health/dwq/gdwq3/en/
- *Manual on Treatment for Small Water Supply Systems*, WRc-NSF, 2001. Aimed at environmental health professionals. Contains good diagrams of water supply systems
- *Private Water Supply Regulations, 1991*. Available from HMSO Publications, or on the Drinking Water Inspectorate website (www.dwi.gov.uk).
- Private Water Supply (Scotland) Regulations, 2006 – guidance on the new regulations for Scotland, and will incorporate guidance for England, Wales and Northern Ireland when the new regulations come into force. Includes the definitive 600 page guidance document on compliance, and a more user-friendly version for members of the public. www.privatewatersupplies.gov.uk
- Drinking Water Inspectorate leaflets are available on the most common areas of concern with mains water supplies. They are downloadable from the website (www.dwi.gov.uk/pubs/index01.htm), or available by post
- *Water Treatment Manual* (Oxfam Water Supply Scheme for Emergencies) – downloadable from the Oxfam website (www.oxfam.org.uk)

- *Water Filtration Manual* (Oxfam Water Supply Scheme for Emergencies) – downloadable from the Oxfam website (www.oxfam.org.uk)

Chapter 5. Rainwater Harvesting

What to use rainwater for – how to use it – parts of the system – doing the calculations – how sustainable is it?

In most parts of the UK, the water that falls on the roof of your house is collected in gutters, directed into down pipes and then gets discharged into the drains as if it is a waste product. This causes a range of problems (discussed on page 30), and you could look upon this rainwater as a resource that you can collect and use for your own purposes. After all, it's clean, free and relatively straightforward. This chapter describes how and why you might do this. Much of the detail on storing, moving and cleaning water that has been discussed in Chapters Two to Four in the context of private groundwater or surface water supplies, applies equally to rainwater, but is repeated here for the benefit of those readers with mains water supplies, so this chapter may be read in isolation.

What can rainwater be used for?

It is technically possible to meet all your domestic water needs using rainwater, but it is rarely an environmentally friendly option in the UK. This is in contrast to many other countries; isolated properties in Australia routinely rely on rainwater for all their domestic needs, largely due to lack of other water supplies and a much sparser availability of mains water supplies. A UK example of a water-autonomous house is described in the case study below.

The self-sufficient house

The house was built as a self-sufficient town house according to the specifications of eco-architects Robert and Brenda Vale and is located in Southwell, Nottinghamshire, where the annual rainfall is 0.6 metres a year. The total roof area collected from is 140m^2 and is made of tile and glass. Water from the roof passes through a coarse filter to storage tanks in the basement that have a capacity of 28,500 litres (i.e. 28.5m^3). The water from these tanks is pumped into a sand filter and from there into an additional storage tank. From this tank the water is pumped to a 250 litre header tank in the loft, which feeds appliances. A spur from this pipe pumps water through activated carbon filters to supply the drinking water needs of the house. Given the limited water supply, a compost toilet is used instead of a flush toilet. The system has been working for 10 years and there has never been a shortage of water, in part due to the exceptionally low water consumption of the residents (around 35 litres/person/day).

How you use your rainwater depends on how much you are prepared to store and how much trouble and expense you are willing to go to, to clean it. The simplest measure is to install a water butt at the bottom of each rainwater downpipe and collect untreated rainwater for use in the garden. This very basic level of rainwater harvesting is probably as far as most people with mains water supplies can and should go. Garden water use is discussed in detail in Chapter Seven. If you collect more rainwater than you can use in the garden, the next priority is to use it for non-potable purposes within the home, so that you can reuse it without concerning yourself too much with cleaning. Referring to Figure 1.7, the major non-potable water demand in the home is the toilet, so this would be the most obvious appliance to use rainwater for. It may also be appropriate to use your rainwater supply to run your washing machine, particularly if your main supply of water is very hard. In practice, if you are using rainwater for both toilet flushing and garden use (the most common scenario), you may be able to meet most of your toilet flushing demand in the winter, but little or none in the summer, when the tank will either be empty, or you will be using the water in the garden.

How to use rainwater

To design a rainwater harvesting system, you will need to calculate supply and demand, and this will be considered on page 122. The ratio of supply to demand will determine how much water you need to store. Rainwater supplies are intermittent: it doesn't rain every day. Your demand for toilet flush water, on the other hand, is likely to remain fairly constant over the course of the year, and demand for garden water will be greatest during the dryer summer months, so you will have to store large volumes if you wish it to meet your entire household demand.

System configurations for a rainwater harvesting system

A rainwater harvesting system generally has the following components; collection surfaces and gutters, filter, storage, pump, pipe-work, cistern/header tank, control system. There are three basic layouts for these components, illustrated in Figures 5.1, 5.2 and 5.3. The relative merits of the three system configurations are indicated in Table 5.1.

Component details

Collection surfaces

When we consider the surfaces we can collect rainwater from, we are concerned with the effects that they have on water quantity and water quality.

Figure 5.1 Indirectly pumped rainwater harvesting system. Rainwater passes through a filter (a) and into the storage tank (b). Water is pumped (c) into a header tank (d), from where it flows to the appliances (e). Mains water backup (f) is provided to the header tank via a suitable air-gap.

Area of collection

The most obvious place to collect rainwater from is the roof of your house, but if you have other substantial roof areas close to the house, collecting from those might also be appropriate. In single-household schemes, collecting rainwater from the ground is rarely appropriate, since it will require additional cleaning unless the water is just used for garden watering. In order to estimate how much rainwater you can collect from your roof, you need to measure its area. This is the amount of ground that the roof covers, not the surface area of the roofing material itself (which will be larger if you have a sloping roof, see Figure 5.4). You should try to avoid collecting rainwater from roofs that have trees overhanging them; either remove the

Figure 5.2 Directly pumped rainwater system. Water passes through the filter (a) and into the storage tank (b). It is then pumped (c) directly to the appliances (d). Mains water backup (e) is to the storage tank via a suitable air-gap.

trees or don't collect the water; it makes more sense to remove the contaminants at source than to try and clean the water once it has already become dirty.

Material and pitch

The slope and construction of your roof influence the run off coefficient: the proportion of rainwater landing on the roof that is actually available for use, compared to that which evaporates. Smooth materials, such as glass, slate, glazed tiles and metal have good run off coefficients and little contamination will be picked up by the water. Turf roofs or sedum are not recommended for rainwater harvesting systems if you are reusing the water inside the house, since the rainwater may become coloured and will collect grit and silt that can corrode your appliances. Rough materials, such as asbestos, felt or wood will tend to collect debris, which

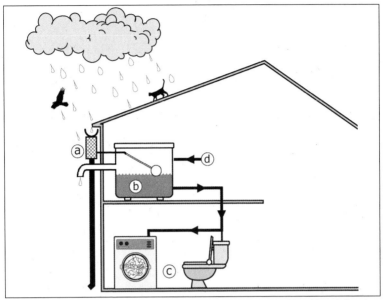

Figure 5.3 Gravity fed rainwater system. Water passes through the filter (a) and directly into a roof-level storage tank (b), from where it flows into the appliances (c). Mains water backup (d) is to the storage tank via a suitable air-gap.

	Indirectly pumped (Figure 5.1)	Directly pumped (Figure 5.2)	Gravity fed (Figure 5.3)
Advantages	Generally simplest system to integrate into existing buildings.	High pressure at appliances so rapid fill rate. Avoids need for separate header tank.	No energy requirement for pumps. Simple and low cost.
Disadvantages	May incur a higher energy cost since rainwater is pumped to a higher level than the appliance it is supplying. Low water pressure at appliances.	No water to appliances during power cuts or pump failure. Backup to ground level tank incurs energy cost as mains water will require pumping to appliances.	Low water pessure at appliances, which will therefore fill slowly. Washing machines may require a higher water pressure. Water may be of lower quality as stored in warm loft. Requires a roof void with weight-bearing capacity for a large water tank. More suited to agricultural than residential buildings.

Table 5.1 Advantages and disadvantages of possible rainwater harvesting system configurations.

Figure 5.4 The amount of rainwater you can harvest is determined by the ground area covered by the roof, not its surface area.

itself will contaminate the water; organic debris such as moss and lichen can be a particular problem as it then provides a food source for organisms in the rainwater storage tank. Wood may also impart colour to the rainwater from tannins.

The steeper the roof, the better the run off coefficient will be. Flat roofs will lose as much as 50% of water to evaporation, particularly during brief rainfall events. The run off coefficient is also influenced by the pattern of rainfall events; less water will be lost to evaporation during sudden showers than the same amount of rain falling as a long period of drizzle.

Rainwater is slightly acidic and will dissolve small amounts of lead roof flashings. Assuming you are not going to drink the water, this isn't a problem, although lead gutters and downpipes should be avoided as they may result in lead levels being too high for rainwater to be used on edible crops in the garden.

Rainwater goods – moving the water from the roof to storage

Gutters should have a 1-2% fall wherever possible, and there should be sufficient brackets to prevent any sag, since standing water will tend to become contaminated by any solids that collect in the gutters. As with roofing materials, smooth surfaces are preferable. All rainwater goods should be easily accessible to allow cleaning.

Some older houses have the waste water from baths and sinks entering the rainwater goods rather than into the soil stack (Figure 5.5), and it is essential to disconnect these (and connect them to the foul water drain) to prevent rainwater being contaminated by this much dirtier water. To collect rainwater from both sides of your roof, it may be necessary to modify the layout of downpipes, but you should make sure that gutters are not overloaded. You can calculate the necessary gutter sizes and number of downpipes using the guidance in the *BRE Good Building Guide 38, Disposing of Rainwater*, or in Part H of Building Regulations (see Further Reading).

First flush mechanisms

The first few millimetres of rain during a rainfall event will pick up the majority of the contaminants from the roof and gutters and effectively wash the roof;

Figure 5.5 If waste water pipes from showers, washbasins etc are connected into your rainwater goods, you will need to divert them into your foul drainage system before making use of your rainwater.

later rainfall will be much cleaner. It is possible to get various types of diverting mechanisms that automatically prevent the first few litres of water from being collected into the tank, thereby improving the overall quality of the water collected. These are most appropriate in situations where the rainfall pattern is of sudden showers, rather than continual drizzle, otherwise the percentage of water collected is considerably reduced (on average the devices reject the first 40 litres or so of water per rainfall event). They are widely used in both Australia and the USA, but are redundant when good quality filters are used and are rarely used in the UK.

Cleaning the water

Whilst it's in the sky, rainwater is very clean. It may contain a few dissolved substances, including air-borne chemical pollutants, but when it hits your roof or the ground, it immediately has the potential to become contaminated by:

- leaves, moss etc
- dust and grit
- bacterial contamination including:
    Salmonella from birds
    faeces from small mammals e.g. squirrels

If collected at ground level, there is the potential for contamination by hydrocarbons such as oil, petrol and diesel, and a considerably increased load of dirt, silt and animal faeces. Consequently, collection from paved surfaces is unlikely to be appropriate, unless water is required in such large quantities that extensive treatment is a viable option, such as in community scale schemes.

With suitable roof and gutter materials and good housekeeping, simple filtration provides sufficient cleaning for the water to be used directly for toilet flushing without any disinfection. Filters can be positioned in the downpipe or underground, and which is the most economic option will depend upon how many downpipes water is being collected from. A typical downpipe filter is shown in Figure 5.6. Collection efficiency can be around 95% (when clean), but will fall to below 30%

if the filter isn't cleaned regularly (algal growth blocks the pores in the filter). If rainwater is to be used for anything other than non-potable uses such as the toilet, it is essential to clean the water as described in Chapter Four.

Storage

Even if the incoming water is clean, contamination problems can be caused by poor storage tank design. For example frogs, midges, rodents and worms may all get into your tank if it isn't properly sealed at the inlet, outlet, lid and overflows. Any pathogens in the water have the potential to breed, particularly if there is any organic material contamination to provide a food source. Storage tanks for rainwater should include the same design features as those for potable water, as illustrated in Figure 3.1, page 52. Tanks are best situated underground and in the dark, where possible, as the lower light and temperature levels will reduce the potential for pathogen growth, although putting the tank underground can complicate the installation of drains and overflows, particularly in existing properties. A range of underground tanks are available, including plastic, fibreglass and concrete. It is possible to have rainwater tanks in the basement, although access difficulties may limit the choice of tank.

Pumps, pipes and controls

Following storage you will need to pump rainwater back into the house, unless you have a gravity fed system (Figure 5.3). Numerous pumps are suitable for rainwater harvesting applications and you should obtain specialist advice from manufacturers before choosing one, as the choice of pump is generally determined by the nature of the control unit. Influencing factors include:

a) is the pump going to be submersible (i.e. in the tank), or suction? Submersible pumps are more efficient and generally quieter

b) how high do you need to pump the water and at what flow rate? This determines the size of the pump you will need

c) how is the pump going to be controlled? It will need to be protected from pumping when dry, but may also be controlled by pressure or the level of water in a header tank

Regardless of the final choice of pump, the inlet pipe for the pump should be suspended about 20 centimetres from the bottom of the tank to prevent any settled solids from the bottom of the tank being stirred up. A fine mesh filter should cover the inlet, since even fine solids will cause considerable pump wear. A suitable inlet can be made out of filter cloth and a float from a ball valve, or proprietary models are available (Figure 5.7).

Figure 5.6 A Wisy downpipe filter. These filters replace a section of your rainwater downpipe and divert the water into a storage tank. When water flows down a vertical pipe, it tends to stick to the edges of the pipe by surface tension. In the filter, a section of downpipe is removed and replaced by a mesh, as shown in the cutaway section. Water passes through the mesh and through the outlet to the rainwater tank.

Back-up systems

It is assumed that a) you will have an additional water supply (be it private or mains) that you can use to flush your toilets during periods of low rainfall, and b) you will want to have this back-up system acting automatically, without you having to check levels in tanks or operate valves. In most cases this back-up supply will be your potable water, and with an automatic back-up system you may be at risk of contamination of your potable water by the rainwater. You should therefore use a failsafe measure to prevent rainwater flowing back into your potable water system as described below. An automated system is likely to be safer, since once installed there is no risk of human error, which might arise if you were relying on regular manual interventions.

The two obvious points at which you could connect your mains water back-up into your rainwater system are directly to the header tank feeding the toilet cistern (Figure 5.1), or to the rainwater storage tank at ground level if your toilet cisterns are currently fed directly from the rising water main in the house (Figure 5.2).

Safety and regulation

Water quality

There are no mandatory national standards for the cleanliness of water used for toilet flushing in the UK, although BSRIA guidance is that it should meet regulations on bathing water quality (see Chapter Four for discussion on water quality requirements). One of the reasons that there is no mandatory standard

Figure 5.7 Inlet pipe for rainwater pump. Water enters the pipe (a) via the mesh filter (b). The position of the filter is kept close to the surface of the water by the float (c) which helps prevent both floating scum and settled solids from entering the pipe.

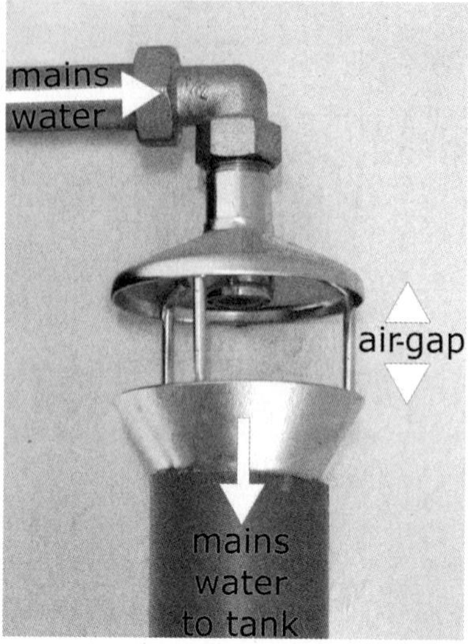

Figure 5.8 Tundish. This device is one way of providing the necessary air-gap to prevent backflow of rainwater into your drinking water supply. Mains water falls through the air-gap into the collection pipe below, and from there to a storage tank. If rainwater were to fill the collection pipe, it would overflow from the tundish rather than enter the mains water supply.

is that the major safety concern with using rainwater for toilet flushing is not related to the potential dangers of water when it is in the toilet cistern or bowl (an already pathogen laden environment), rather it is to do with the risk of cross-contamination of other parts of the water supply system. Increasing numbers of systems, particularly larger scale installations, are including UV disinfection. This may negate the environmental benefits of using rainwater in the first place, and since UV is only completely effective on water with no solid content it is of limited usefulness for rainwater that has only passed through a coarse filter.

Given that the rainwater you are using has not been treated to drinking water standards, it is essential that precautions are taken to prevent cross-contamination of the rest of your water supply or the public mains. There are two basic sets of precautions that must be taken to minimise this risk.

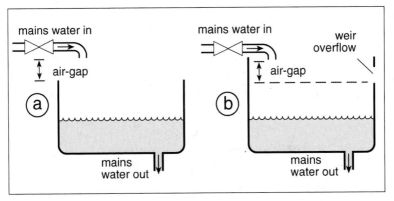

Figure 5.9 (a) Type AA air-gap with an unrestricted discharge. (b) Type AB air-gap with a weir overflow. The air-gap must be 20mm or twice the diameter of the inlet pipe's bore diameter, whichever is greater.

Labelling

The Water Regulations Advisory Service (WRAS) publishes guidance on labelling of reclaimed water systems. This guidance should be referred to during installation of any system, and basically requires all pipes containing reclaimed water to be labelled in a specific way, to prevent inadvertent connection of reclaimed water pipes with potable supply pipes. Labelling of reclaimed water supply points (taps etc) is also required to indicate that the water is not suitable for drinking.

Backflow

Under the Water Supply (Water Fittings) Regulations, 1999, varying degrees of precaution must be taken against backflow, depending on the degree of contamination of the water. Contaminated water is therefore divided into a number of fluid categories. Rainwater falls into Fluid Category 5 (since it may constitute a serious health hazard owing to the likelihood that it contains pathogens). The regulations therefore require the fitting of a 'Type AA, AB, AD or AUK1 air-gap'. This is not as complicated as it sounds! An example of an air-gap that fulfils the requirements of the regulations is shown in Figure 5.8, and you should check with the supplier that the fitting constitutes a suitable air-gap under the regulations (as shown in Figure 5.9).

Informing the water supplier

Under the Water Supply (Water Fittings) Regulations, you must tell your water supplier that you are installing a rainwater harvesting system. They may wish to inspect your installation, and at the very least they will want to make sure that you have followed precautions to avoid contamination. They may also wish to check the provision you have made for overflow from the rainwater storage tank, since you are not allowed to discharge rainwater into the mains sewer.

If you are on a private water supply and have no mains water connection, you are not obliged to comply with these regulations since there is no risk of contamination of the public mains water supply. However, they do represent best practice and given the potential health consequences of cross-contamination we would strongly advise you to follow them.

Doing the numbers

The major variable between rainwater harvesting systems in different properties will be the size of the storage tank. If you are purchasing a rainwater harvesting system, the vendor will usually calculate your tank volume for you. We would recommend checking what calculation method will be used, as different vendors may use different methods and you may be able to save yourself a lot of money by opting for a slightly smaller tank, without significantly decreasing the water demand that is met by rainwater. Two methods for calculating storage volume will be described here, the first based simply on annual rainwater supply, the second using a model that includes both supply and demand. The two methods are compared using the example in the box on page 125.

Regardless of which method you use, the amount of rainwater you will have available for your use depends upon the following factors:

- rainfall
- collection area
- nature of collection surface
- efficiency of any cleaning systems

Rainfall: The national average rainfall in the UK is almost exactly 1 metre/year, but this average hides a lot of variation as shown in Figure 5.10. Even within very small geographical areas, rainfall can vary dramatically due to land topography. Rainfall can also be highly seasonal, so you will have to size your tanks carefully if you expect your rainwater system to meet your demand during the summer, particularly if this demand will include garden watering. It is usually worth getting daily or monthly rainfall figures so you can more accurately calculate a tank size, rather than overestimate what is required and find that your tank is rarely full. Rainfall data from your nearest weather station is available from the Meteorological Office, although you will have to pay for it, so it is worth ascertaining what data the supplier of your rainwater harvesting system already has.

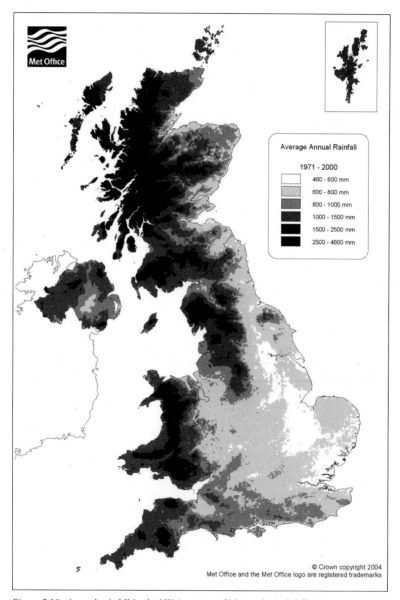

Figure 5.10 Annual rainfall in the UK (courtesy of Meteorological Office).

Material and pitch	Run off coefficient
Roof tiles at average pitch	0.75-0.9
Flat roof with a smooth surface	0.5
Flat roof with gravel, turf or sedum	0.4-0.5

Table 5.2 Run off coefficients for various roof materials and pitches.

Run off coefficient: The roof material and how sloped it is will affect how much of the water that falls on the roof is actually collected. Typical run off coefficients are given in Table 5.2.

Filter coefficient: Depending on the type of filter you use, and how often you clean it, you can expect filter efficiencies to be over 80%. Most are designed to maintain this efficiency over a wide range of flow rates and you should refer to the manufacturer's data for a filter coefficient.

Method 1 – calculation of storage volume based on supply of rainwater

The basic equation you need to calculate your annual water supply is:

$$\frac{\text{Yield}}{\text{m}^3/\text{year}} = \frac{\text{annual rainfall} \times \text{collection area} \times \text{run off coefficient} \times \text{filter efficiency}}{\text{metres} \qquad \text{m}^2}$$

The simplest rule of thumb is to have a tank that can store 5% of your annual rainwater supply, i.e. 18 days supply. You should choose the next available tank size above that indicated by this result, since the volume of rainwater available for use from the tank is lower than the actual volume (taken to the level of the overflow), and since there will be standing water in the bottom of the tank that is not used.

Method 2 – storage volume based on demand and rainfall data

Practical experience of rainwater harvesting in the UK has found that if calculated according to supply only (as in method 1), the rainwater tank is rarely full, which begs the question of how much smaller a tank could be and still meet an acceptable demand. This in turn affects tank siting and financial payback. A more accurate way of calculating the appropriate storage volume is to base it on a combination of supply and demand. Computer programs are available that will allow you to do this. One such model is available from the Environment Agency (and can be downloaded from their website as an Excel spreadsheet; follow links to the 'Waterwise on the Farm' project). Whilst it is designed for use by farmers, to calculate the necessary storage volumes for agricultural rainwater use, it is a reasonable guide for domestic rainwater harvesting system calculations. To use this model you will need to know your rainfall data in as much detail as possible (the spreadsheet allows input of yearly, monthly or daily rainfall data, and more robust results will be achieved if monthly or daily data is used rather than annual). Additionally, you will need to input your collection area, your roof run off coefficient, and your filter coefficient as in method 1.

You will also need to calculate your water demand, and what follows assumes that you will be using the water for toilet flushing only. There are a number of ways in which you can calculate your demand for toilet flushing:

a) 35% of annual water use, taken from water bill if you are on a metered supply

b) Approximately 50 litres/person/day, assuming a 'normal' 9 litre toilet

c) 30 litres/person/day, assuming use of an ultra low flush toilet

d) Measure directly over a period of weeks using a water meter on the toilet water supply pipe

The model is fairly straightforward to use; you simply input the necessary data in the cells indicated, and the spreadsheet then calculates a graph indicating the percentage of water supply collected and demand met over a range of tank volumes.

Comparison of the two methods

The two methods are compared in the box below. You will see that the recommended tank size is smaller using the supply and demand based approach (method 2), and larger tanks will have progressively less and less impact on the total demand met.

Case Study – Calculating tank sizes for rainwater harvesting systems ——————

Supply: A house in Machynlleth has an available roof area for rainwater collection of 50m². The roof is slate and has a steep pitch: the run off coefficient is therefore taken as 0.8. A filter coefficient of 0.8 is used, and annual rainfall is 2.19 metres. The annual water supply is therefore 2.19 x 50 x 0.8 x 0.8 = 70.1m³

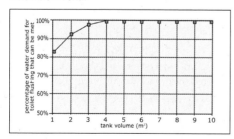

Figure 5.11 Percentage of water demand met by rainwater at various tank sizes.

Demand: There are 4 occupants, none of whom are at home during the day. A dual flush toilet is installed, and water use for toilet flushing is estimated at 120 litres/day.
Tank size according to method 1: 5% of 70.1m^3 is 3.5m^3.
Tank size according to method 2: The graph of tank size versus percentage of demand met using daily rainfall figures is shown in Figure 5.11. You can see from this that a 1m^3 tank will allow rainfall to meet 83% of demand, and there is progressively less benefit as tank size increases.

As is always the case with models, rainwater harvesting tank sizing guides are only as good as the assumptions they make and the data that you put into them. An independent study into the payback of rainwater harvesting systems demonstrated that most current models are fairly inaccurate due to use of insufficient data and over-simplistic treatment of supply and demand. As a result of ignoring real running costs, many are also over-optimistic about paybacks (see Further Reading for details).

Rainwater harvesting and sustainability

Whether or not rainwater harvesting is more or less sustainable than using mains water and drainage, is very much dependent on your individual situation (a variety of which are discussed below). Whatever your situation, water efficiency measures (as discussed in Chapter One) and collecting as much water for the garden as you can use (as discussed in Chapter Seven) are higher priorities than installing a rainwater harvesting system. Since much of the environmental cost of water supply is in the installation of infrastructure, rainwater harvesting systems are unlikely to be sustainable in terms of their impact on total water demand, unless carried out on a very large scale so that less central infrastructure is necessary. In terms of the continued environmental costs of the cleaning and supply of water, areas with a scarce water resource are also those with low rainfall (compare Figures 1.1, 1.2 and 1.3 with Figure 5.10). Therefore even widespread uptake of these systems would have little impact on total consumer demand in the summer, since rainwater storage tanks would be empty. There are occasions when combination systems with other water sources (e.g. grey water, as discussed in Chapter Six) are appropriate.

Environmental sustainability

The most detailed analysis to date of the environmental sustainability of rainwater harvesting systems has used the technique of Life Cycle Analysis (LCA). A Life Cycle Analysis comparison of rainwater harvesting systems with mains water supply when used for toilet flushing in Switzerland, suggested that the environmental costs of rainwater harvesting marginally outweighed the benefits in domestic situations. However, the boundary conditions in this study did not include any costs associated with conventional rainwater disposal infrastructure, and disposal to soakaway was assumed. As discussed on page 30, the rainwater disposal infrastructure in the UK is far from sustainable, and if these environmental costs were included in an LCA it may well be that rainwater harvesting begins to look like a more sustainable option. Choice of tank material is also critical; more concrete is required to backfill an excavation for a fibreglass tank than would have been required for installing a concrete tank in the first place. Polyethylene tanks on the other hand do not usually require the excavation to be backfilled with concrete. Risk-averse engineers can also negate the benefit of rainwater harvesting systems by specifying UV disinfection for the rainwater. A major determining factor in the sustainability of rainwater harvesting systems is what would otherwise happen to your rainwater and surface water. If it would go to a mains drainage system, then rainwater harvesting may make sense. However, if rainwater would otherwise be discharged to a soakaway as is common in rural areas, it will recharge the local groundwater and it is probably better to collect water for garden use and allow the remainder to discharge to soakaway than to install a rainwater harvesting system.

New build urban developments

When planning permission is granted for new housing developments in urban areas, it often includes a stipulation of maximum rain and surface water run off. The development may therefore need a sustainable urban drainage system (SUDS) in order to slow down the flow of surface water from the property and help prevent mains drains becoming overloaded. A common way of slowing down the flow of surface water in these developments is to use attenuation tanks, which collect the rainwater and then release it more slowly into the drainage system. In such a development, where tanks are required as part of the design, rainwater harvesting may well make sense, both environmentally and economically. However, in most instances installation of rainwater harvesting systems do not negate the need for SUDS, as it is difficult to optimise designs for both rainwater reuse and attenuation capacity.

Existing properties, urban and rural

Rainwater harvesting systems are typically more difficult and more expensive to retrofit into properties than to install as part of the initial build process. A professionally installed system will cost in the region of £2000-£3000, and you should expect to have to replace some of the components after around 5-10 years. The running costs of a system may well outweigh the savings on mains water. Your first priority should be water efficiency, followed by collecting as much rainwater as you can use for garden watering. In rural areas, you can allow the remainder to infiltrate into the ground via a soakaway. In the absence of mains drainage, rainwater harvesting makes little sense either environmentally or economically; whether rainwater harvesting is more sustainable than using mains water is critically dependent on the unsustainability of the mains drainage system. Removing the need for toilet flushing by installing a compost toilet would be a better environmental option than using rainwater for flushing.

Non-domestic properties

The situation in which rainwater harvesting has been used with most success in the UK, is in non-domestic buildings in urban areas. These buildings often have a large roof area, and since they are only used during the daytime, a high proportion of their water demand is for toilet flushing (Figure 1.7). These buildings are also likely to be on metered water supplies, and may also be charged for disposal of rainwater in addition to sewage. Specific industrial applications that have a high demand for water that doesn't need to be of potable standard (e.g. car washes) are also well suited to rainwater harvesting.

Rainwater harvesting may also be a reasonable proposition for community scale housing schemes, as discussed in the box below.

A community scale water strategy

This development of 76 flats and houses for Acton Housing Association was granted planning permission subject to the developers minimising the amount of storm water run off from the site, and the amount of potable water used in the dwellings.

The strategy is to collect rainwater from all the roofs on the development (approx $2250m^2$), filter it and then store it in tanks around the site with a total capacity of $59m^3$. From these tanks, the rainwater is fed back into 15 of the properties and is used for flushing toilets. The impact on the mains drainage system is to prevent over $870m^3$ of rainwater a year from entering the storm water drains.

In the properties supplied with rainwater, the tenants will save £64 a year on their water bills. The rainwater harvesting systems are estimated to supply 80% of the flushing requirement for a 5 person household. To minimise the amount of mains water used on the site as a whole, all houses are fitted with ultra low flush, dual flush Ido WCs and aerated basin taps. This translates into a saving of $1573m^3$ of mains water per year from the WCs alone.

Rainwater from hard standings on the site is either run off to adjacent soft landscaped areas or runs into the storm water drains. Permeable paving was suggested as part of the overall strategy but did not make the final design.

In terms of sustainability, for most domestic applications in which rainwater would be used for toilet flushing, it would be more sustainable to install a compost toilet and remove the need for toilet flushing entirely than it would be to install a rainwater harvesting system. In the future, rainwater harvesting may have an important role as part of a more integrated approach to water resource management, particularly when carried out on a community scale. However, at an individual level, there are better ways of greening your life than by investing in a rainwater harvesting system.

Summary

It is rarely appropriate to try and meet your entire domestic water needs with rainwater in the UK. If you are going to collect rainwater, first implement water efficiency measures, with rainwater collection for the garden being the next priority. There are few domestic situations in which harvesting rainwater to use for toilet flushing is appropriate; it is a technology more suited to community scale development and commercial buildings. Careful attention to detailing of collection surfaces, gutters, filters and storage tanks allows collected rainwater to be of good quality, and calculations of rainwater yield should be considered in light of water demand in the building.

Further Reading

- *Water Supply (Water Fittings) Regulations*, 1999 – regulations that include means of preventing backflow from rainwater systems into drinking water systems
- Water Regulations Guide, WRAS, ISBN 0-9539708-0-9 – guidance from WRAS (Water Regulations Advisory Scheme) on the Water Supply (Water Fittings) Regulations 1999
- *Reclaimed Water Systems* – information about installing, modifying or maintaining reclaimed water systems. WRAS information and guidance note 09-02-04
- *Waterwise on the Farm* – guidance on rainwater collection available from the Environment Agency website (www.environment-agency.gov.uk). Predominantly aimed at farmers, but includes a useful model that you can use to calculate rainwater storage volumes
- Harvesting rainwater for domestic uses: an information guide, (Environment Agency, 2003)
- Predicting the hydraulic and life-cycle cost performance of rainwater harvesting systems using a computer based modelling tool. Roebuck & Ashley, 2006. Paper presented at the 7th International Conference on Urban Drainage Modelling. Available to download from www.SUDSolutions.com
- *Rainharvesting in Buildings, The Rainwater Technology Handbook*, Klaus W Konig, Wilo-Brain, ISBN 3-00-008368-3, 2001 – details of a number of

rainwater harvesting projects from around the world

- *Approved Document H, Drainage and Waste Disposal*, The Building Regulations, 2000 – how to calculate roof gutter size according to rainfall and roof area
- *Disposing of Rainwater*, BRE Good Building Guide 38, ISBN 1-86081-382-8 – how to calculate roof gutter size, with useful advice on design and installation of gutters
- *Life cycle assessment of drinking water and rainwater for toilet flushing*, Crettaz P., Jolliet O., Cuanillon J.M. & Orlando S., Aqua 48:3, pp. 73-83, 1999
- *Rainwater and grey water use in buildings: decision making for water conservation*, D. J. Legget, R. Brown, G. Stanfield, D. Brewer & E. Holliday, CIRIA PR 80, 2001, ISBN 0-86017-880-3 – discusses the background to the use of rainwater and grey water, the first of three publications that resulted from the 'Buildings that Save Water Project' (BTSW) undertaken by CIRIA and BSRIA.
- *Rainwater and grey water use in buildings, Best Practice Guidance*, D. J. Legget, R. Brown, G. Stanfield, D. Brewer & E. Holliday, CIRIA C539, ISBN 0-86017-539-1, 2001 – detailed guidance on design principles for rain and grey water systems
- *Rainwater and grey water in buildings: project report and case studies*, Technical Note TN 7/2001, D. Brewer, R. Brown & G. Stanfield, ISBN 0-86022-577-1 – case studies of a variety of rain and grey water systems in the UK within the 'Buildings that Save Water' project
- Meteorological Office (www.met-office.gov.uk) has rainfall data for the UK

Chapter 6. Grey Water Recycling Systems

Why recycling your water isn't necessarily a good idea – what water might you recycle? – how might you recycle it? – combining sources of recycled water.

The preceding chapters have suggested that we are going to have to go to quite considerable trouble to find our water supply, to clean it prior to use and to distribute it around the home. You might conclude from this that we should use it as many times as possible, before passing it back into the water cycle. You may have heard about grey water recycling systems and be aiming towards a completely 'closed loop' where you treat your dirty water from toilets, showers etc and make it clean enough to reuse. It is technically possible to do this, but you should certainly not consider this to be some ecological Holy Grail; rather the opposite, as you will see below; in the UK recycling domestic water for reuse within the house is rarely environmentally friendly.

You will remember from Chapter One that the first thing you should do if you are aiming for a sustainable water system is to implement water efficiency measures. One result of this will be to reduce your dirty water production, and therefore make it harder to justify a complicated recycling system. If you are a gardener, you could install a simple recycling system for reuse of household water, discussed in Chapter Seven.

Recycling water you have already used is not appropriate if you have a mains supply or access to a spring, well/borehole, surface water or rainwater supply, as these will probably be both more plentiful and of better quality. The type of sewage system you have also has a bearing on water recycling. If you are not on mains sewerage and are about to build your own system, then you could design it to include the potential for reuse of the final treated effluent.

But what if there really isn't any other available water source? If you are in a situation of extreme water scarcity, where your primary water source is insufficient to use for anything other than drinking, cooking and washing, then you may be considering recycling for some of your non-potable water uses. You would still do better to minimise these non-potable uses by installing a compost toilet.

If you are still determined to recycle water for reuse in the home, the remainder of this chapter looks at how best to do it.

Which water sources are worth recycling?

With all water sources you have to consider the following factors:

- amount available (and how regular this amount is)
- quality (potential to contain contaminants)
- type of use to which the water is going to be put (toilet flushing, garden irrigation or all domestic purposes)
- ease of getting the water

As you will see from Table 6.1, the least dirty and most abundant of these 'waste' water streams in the home is the water arising from baths, showers and washbasins.

Source of water	% total domestic water use (see Figure 1.7)	Regularity of production	Degree of contamination	Other points to consider
Toilet	35	Very regular	Very high	Compost toilet is likely to be more appropriate than recycling toilet flush water.
Washing machine	12	Variable depending on household	Wash: high Rinse: low	Difficult to separate machine wash water from rinse water. Washing nappies will significantly increase pathogen load.
Kitchen sink, dishwasher	19	Very regular	High	Fats and grease cause particular difficulties.
Baths, showers, wash basins	28	Very regular	Low	
Car washing, gardening	6	Irregular	Car washing: High	Collection is difficult as is on ground surface.

Table 6.1 Sources of dirty water in the home and factors affecting their suitability for reuse.

Toilet water is just too 'dirty', water from the kitchen sink contains fats and food, and washing machine water contains high levels of detergents. The water from baths, showers and washbasins is often known as grey water and is potentially contaminated with:

- hair and skin
- fats and grease from washing
- food waste   } nutrients
- dirt and grit
- human faecal matter
- urine
- animal faecal matter (from pets or outdoor workers)   } bacteria and pathogens
- detergents, soaps, surface cleaners

Since grey water contains bacteria and a nutrient source, and is often discharged warm, you have an ideal situation for pathogens to multiply. Studies of installed systems in the UK have demonstrated coliform results in excess of 1000/100 millilitres. Consequently, if you are going to install a grey water system that recycles water for use in the house, it is essential to treat the water and even then use it only for non-potable purposes such as toilet flushing.

Methods for recycling grey water

There are two basic approaches to grey water recycling systems. The first approach is to regard them as sewage treatment systems. The solids must be removed and the liquid filtered through a reed bed, sand filter or similar, as described in *Lifting the Lid* and *Sewage Solutions* (see Further Reading). There is a system available in which this filtering function is carried out by a green roof. The treated water could then be stored in a pond, ready for reuse. This pond water should then be analysed for potential contaminants in the same way that you would do for any other potential water source, and an appropriate cleaning method employed to bring the water up to recommended standards for toilet flushing, as described in Chapter Four. It is worth reiterating once again, that it would be preferable to reuse this pond water in the garden, and/or to install a compost toilet than to bring the treated water back into the house.

The second approach to grey water recycling is to buy a system 'off the shelf'. Given their complexity, it is assumed that if the reader wishes to pursue this avenue, they will have a system professionally installed. A typical example of such a system is described below.

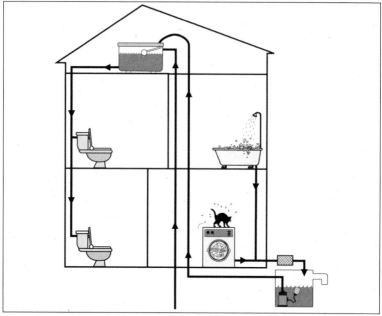

Figure 6.1 Grey water recycling system. Grey water from showers, baths, washbasins and washing machines is collected in a storage tank and cleaned before being pumped to a header tank, from where it is distributed to the toilets. Mains water back-up is provided to the header tank via a suitable air-gap.

Off the shelf systems

System configuration

Systems usually consist of a storage tank, filtration system, chemical dosing system, a pump, header tank and some connecting pipe-work. A typical layout is indicated in Figure 6.1.

Cleaning the water

Grey water is filtered when entering the tank, and the filter is usually cleaned automatically by backwashing, with the backwash water discharged to a drain (which will inevitably reduce the amount of grey water available for reuse). The water is then disinfected with bromine. chlorine, ozone or UV. The concentration of disinfectant is designed to produce complete pathogen kill in a variety of circumstances, and levels are sufficiently high to impair the essential microbiological action in private sewage treatment systems (they are not ideal for mains sewage systems either, but dilution minimises the effects).

Storage

As indicated in Table 6.1, the amount of grey water produced by a household (i.e. the amount of clean water used in showers, baths and wash basins, about 28% of total domestic water use) is similar to the amount of water required for toilet flushing (approximately 35% of total water use). In other words, the supply to demand ratio is good, so grey water storage tanks can be fairly small (often around 100 litres for a domestic system) as they are replenished as regularly as they are emptied. The filter and disinfectant dosing system is usually situated in this tank (Figure 6.2).

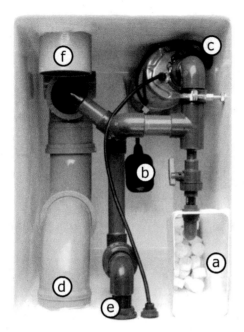

Figure 6.2 Grey water storage tank complete with disinfectant dosing system (a), float switch (b), pump (c). Water enters the tank via pipe (d), and leaves after being filtered and disinfected through pipe (e) Overflow (f) (courtesy of Gramm Enviromental).

Header tank

Treated and disinfected water is pumped from the storage tank to the header tank, which will have a mains water back up, protected from contamination by backflow using air-gaps similar to those in rainwater harvesting systems. In some systems, a combined grey water storage tank and mains water inlet is concealed in a service void behind the toilet (e.g. those manufactured by 'Ecoplay').

Final water quality

The water quality resulting from commercially available grey water recycling systems poses a low risk to health if the system is working correctly. However, more abstract indices of water quality such as odour (e.g. from the disinfectant), colour and cloudiness are often poor. Some systems include disinfectant with a deliberate colour, to provide a visual check for the user that disinfection has occurred, to mask the cloudiness and to distinguish the water from potable supplies. In addition, the high level of bromine may degrade some plastics and rubber found in plumbing systems (e.g. toilet cistern inlet valves).

Safety features

There is a real risk to health with grey water recycling systems if the treatment systems fail, or cross-contamination occurs; consequently they should have a variety of safety mechanisms integrated into them. These can include:

- automatic drain down and discharge to sewer of both the storage tank and the header tank in the event of the system not being used in 24 hours. Whilst good for safety, this reduces the volume of water available for reuse
- disinfection systems that fail in a deliberate way (e.g. prevent the supply of grey water for reuse if the disinfectant runs out), again, good for safety but will reduce the volume available for reuse
- initial disinfection of water entering the tank
- separation of the inlets to the header tank to prevent backflow of grey water into the potable water supply
- artificial colouring to the water to discourage use for other purposes
- pipe labelling to prevent inadvertent cross-connection of grey water into the potable water supply in the house
- an overflow to the sewer that is protected from backflow

Does domestic grey water recycling have a future in the UK?

Grey water systems rarely, if ever, pay back economically; all independently published case studies of installed systems have running costs higher than that of mains water supply. Given the infrastructure requirements and the disinfectant doses needed, grey water recycling systems cannot be considered environmentally friendly, and it is far too complex a system to be appropriate for use by private individuals on a small scale. The value of recycling water with such high coliform counts is extremely dubious from a social perspective, and many people will find the aesthetic water quality in such systems unacceptable. Grey water recycling is still a relatively young technology, and as such may improve with time, but shortages of water in the UK would have to become much more acute to justify the widespread installation of small scale domestic grey water systems. Advances in water treatment technologies (such as the use of membranes, which can produce high quality water even from sewage) may in future allow heavily contaminated water to be cleaned to suitable standards fairly cheaply, although the environmental impacts of this are still high compared to starting with a less contaminated water source. In new build isolated properties where there is no mains water or drainage and other water sources are scarce, some grey water recycling may be appropriate (if implemented as part of a sewage treatment system involving simple filtration (e.g. by a reed bed or sand filter outside the house), rather than using one of the currently available grey water recycling kits described above. In private sewage treatment systems where leachfield overloading is a problem, diverting grey water into a recycling system may improve sewage treatment. Nevertheless, a compost toilet would normally be preferable to recycling grey water for toilet flushing.

Variations on grey water recycling systems; commercial buildings, community scale systems and combined rain/grey water systems

Some light industrial or agricultural enterprises exist where non-potable water use is high and grey water production is high, so grey water recycling may be justified (for example garages with automated car washes, or vegetable washing plants), but most commercial and public buildings are only in daytime use, so grey water production is too low to justify a recycling system. Residential buildings (e.g. universities, hotels) have higher grey water production, but rainwater harvesting may well be more appropriate, since these buildings often have a large roof area compared to the number of building users. Grey water systems serving groups of houses with centralised collection, storage and treatment have been installed at a number of sites in the UK. In these instances, the filtration and disinfection procedures and management regimes are more elaborate than in single-household systems, as the risk to public health from poorly performing systems is greater. Whilst the economies of scale favour these systems over single-household systems, it is still difficult to see how they can be considered sustainable at the present time.

Installing an integrated system that combines rainwater and grey water is also possible. This has the potential to increase the volume of reclaimed water available, with little additional infrastructure cost (at least in theory). Systems can combine the two water sources either before or after treatment, and as indicated in Table 6.2, combination after treatment is likely to be more sustainable.

Combined rainwater and grey water systems require site specific design and are not yet widely available, nor have they proved themselves robust and cost effective either domestically or on a community or commercial scale. An additional problem with these systems is what happens to any overflow from the tank. Since rainwater is not allowed to be diverted to mains sewers, special provision would need to be made for conditions where the combined rainwater/grey water tank was overflowing. Water efficiency measures and an emphasis on storing rainwater are more likely to be appropriate than incorporating grey water in most instances.

Combination before treatment	Combination after treatment
Only one collection tank and pump needed	One tank and pump required for each water type
High environmental and cost impacts of treatment since whole volume must be assumed to be from dirtiest source	Lower treatment impacts
Potentially leads to rainwater overflowing to foul sewer, highly unsustainable	Rain water tank can be allowed to overflow to surface water drainage since it contains no foul water contaminants
Mixing of source water does not allow categorising use according to source	Allows different qualities of water to be used for different purposes

Table 6.2 Combination rainwater and grey water systems.

Summary

Recycling your grey water so that you can use it to flush your toilet is not justifiable on environmental, economic or social grounds. Whilst at first sight it might appear to be efficient to implement a 'closed loop' water recycling system, we cannot consider water in isolation from other environmental problems. Grey water recycling systems typically require energy and disinfectants with higher environmental impacts than can be justified from the amount of water that is saved. If you want to make use of this water, then using it in a simple garden irrigation system, as described in Chapter Seven, is far more sustainable.

Further Reading

- *Lifting the Lid: An ecological approach to toilet systems*, P. Harper & L. Halestrap, CAT Publications, 1999. Available from CAT Mail Order.
- *Sewage Solutions: Answering the call of nature*, Grant, Moodie & Weedon, CAT Publications, 2002. Available from CAT Mail Order.
- *Water recycling at the Millennium Dome*, S. Hills, A. Smith, P. Hardy & R. Birks, Water Science and Technology 43:10, pp287-294, 2001.
- *Rainwater and grey water use in buildings: decision making for water conservation*, D. J. Legget, R. Brown, G. Stanfield, D. Brewer & E. Holliday, CIRIA PR 80, ISBN 0-86017-880-3, 2001 – discusses the background to the use of rainwater and grey water, the first of three publications that resulted from the 'Buildings that Save Water Project' (BTSW) undertaken by CIRIA and BSRIA.

- *Rainwater and grey water use in buildings, Best Practice Guidance*, D. J. Legget, R. Brown, G. Stanfield, D. Brewer & E. Holliday, CIRIA C539, ISBN 0-86017-539-1, 2001 – detailed guidance on design principles for rain and grey water systems.
- *Rainwater and grey water in buildings: project report and case studies*, D. Brewer, R. Brown & G. Stanfield, Technical Note TN 7/2001. ISBN 0-86022-577-1 – case studies of a variety of rain and grey water systems in the UK within the 'Buildings that Save Water' project.
- Selection and evaluation of a new concept of water supply for Iljburg, Amsterdam. van der Hoek, J. P., B. J. Dijkman, G. J. Terpstra, M. J. Uitzinger and M. R. B. van Dillen, *Water Science and Technology*, Volume 39, Issue 5, pp. 33-40, 1999 – study that uses Life Cycle Analysis (LCA) to determine sustainable water supply options for a new development
- *Marking and Identification of Pipework for Reclaimed (grey water) systems*, WRAS information and guidance note, 9-02-05.

Chapter 7. Water in the Garden

How much is used – why to use less - techniques for using less – simple reuse of rainwater – simple reuse of grey water – irrigation system design – coping with hosepipe bans.

Water use outside the house accounts for just 6% of domestic water use in the UK, i.e. 9 litres per person per day, but is increasingly an area of focus for water efficiency campaigns. There are several reasons for this. Firstly, outside water use peaks precisely when and where water supplies are at their most stressed; summer demand for water is a third higher than in winter: this necessitates massive reservoirs to store winter water for use in summer, and is largely attributed to garden watering. Secondly, the use of hosepipes and sprinklers by large numbers of households simultaneously, can result in decreases in supply pressure and subsequent problems in the distribution network; up to 70% of water demand on summer evenings can be due to use in the garden. Thirdly, water use in the garden is increasing, and this is expected to continue due to increasing affluence and interest in gardening, and the effects of climate change (which is expected to lead to wetter winters, but dryer summers in the UK).

Water in plants has a number of roles. It is important for plant structure, takes part in many chemical and physical processes within the plant, and acts as a solvent – allowing plants to take up nutrients. Water is continually flowing through plants: it enters through the roots, travels up through the plant and is then lost through the leaves via a process called *transpiration*. In addition to plants losing water by transpiration, water is also lost directly from the ground by evaporation and the rate at which this occurs will affect how much soil moisture is available to the plant. It is possible to calculate the rate at which evaporation and transpiration occur, but since these rates vary according to species, size, humidity, soil moisture content, soil type, temperature and light levels, it is difficult to water plants in a very efficient way using tables of theoretical transpiration and evaporation rates. Much better is to inspect the soil at a spade's depth. If it feels damp, there is probably no need to water anything. A general guide to watering frequencies is given in Table 7.1.

Making do with less water – water efficiency in the garden

The basic cornerstones of water efficient gardening are to choose the right plants, add plenty of organic material, and use mulches wherever possible.

Water whenever soil seems dry (at a depth of 30cms)	Water every 10-14 days in dry periods	Never water
Perfect lawns	Smaller fruit trees and currant bushes	Everyday lawns
Leafy salads	Vegetables where the fruit is eaten (courgettes, squash etc.)	Wildflower meadows
Peas and beans in flower	Bedding plants	Rough grass
Newly planted items		Large fruit trees
Containers and hanging baskets		Root vegetables, perennial borders and ornamental grasses, drought-resistant succulent plants

Table 7.1 Watering frequency guide courtesy of the Royal Horticultural Society.

Planting scheme

It sounds obvious, but you should try to grow plants that are suited to the local microclimate. If a plant in your garden seems particularly prone to withering away every summer, consider replacing it with something more suitable, or just see what appears. This isn't about limiting your choice of plants, it's more that the plants that are most drought tolerant will look best, as they'll inevitably be healthier than plants that you need to spend a lot of time looking after. Try and rely on plants whose foliage and overall form look good, as this will always look better than wilting flowers. Drought tolerant plants often have pale, silvery leaves, and are sometimes hairy, leathery or waxy. Bear in mind that you will have different microclimates within your garden; some areas will be more shaded from sun and wind, and therefore suffer less from evaporation. If you have favourite plants that don't survive well through the summer, make sure they're planted close to the water butt so that they are easier to water, or put them somewhere where they will be easy to irrigate with grey water. Plant new specimens in autumn or spring so that they have an opportunty to grow a strong root system before a dry summer. Several books consider plants suited to dryer parts of the UK (see page 159).

Organic material

Whether your soil is sand, gravel, chalk or clay, it will be vastly improved by the addition of compost. Organic material will improve the texture of soil; if you have clay soil it will help open up the structure and improve drainage, and, at the other

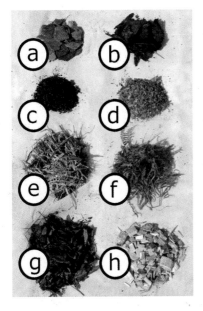

Figure 7.1a (left) Organic mulches.
(a) pebbles (b) bark (c) cocoa (d) sawdust
(e) straw (f) grass (g) leafmould
(h) wood chip.
Figure 7.1b (above) Sheet mulches.
(a) sack (b) geotextile (c) carpet
(d) cardboard.

extreme, if you have a sandy soil it will act as a sponge to retain moisture. In both cases, the organic matter will make it easier for your plants to withstand periods of drought. According to the Royal Horticultural Society, adding compost to soil is equivalent to an extra 5 centimetres of rainfall. You should dig in to the soil as much of your kitchen scrap and garden-waste compost as you can, but if that sounds like too much effort you can simply spread it on the soil surface, or put it in trenches to target it to specific plants. You can also buy in well rotted manure if it is available locally.

Mulches

Mulches are essential to water efficient gardening; they slow evaporation so that plants don't need watering as frequently, and they also suppress weeds (which use water that could otherwise be used by your precious plants). There are two basic types; organic mulches (Figure 7.1a) and sheet mulches (Figure 7.1b). Any organic material that will compost is suitable, such as wood chips, bark, coir and semi-composted kitchen waste. The loose texture of organic mulches means that any weeds that do appear are easier to pull out, and the mulch provides a good habitat for invertebrates (including, unfortunately, slugs!). As the organic mulch breaks down, it will increase nutrient levels in the soil, loosen the texture, and

Figure 7.2 (left) Sinking an empty pot in the earth next to a plant allows you to tell how much water you've given it and reduces the amount that runs off the ground surface.
Figure 7.3 (above) A small depression around each plant minimises the surface water run off when it rains.

improve water-holding ability. Sheet mulches can increase the surface water run off if they are completely impervious, but woven and porous sheet mulches are also available. In terms of sustainability, reused or recycled materials are preferable, such as old carpets (not the foam backed ones), or cardboard (since this will eventually rot down to compost, it is a cross between an organic and a sheet mulch). Unfortunately, these often look messy, so you may want to put organic mulch on top of the sheet mulch for aesthetic reasons.

Watering and gardening techniques

1. If you need to use a hose to water plants in different areas of the garden, fit a trigger nozzle so you're not wasting water whilst moving from one area to another.

2. Avoid watering little and often, as this will encourage roots to the surface where they will dry out easily. Watering every seven to ten days is preferable (and less work). You will have heard this mantra before, but it is not as easy to implement as it sounds, because the water won't soak into the ground very rapidly, so it can be quite time consuming to give a plant enough water without most of the water running off the surface. The following techniques may help.

3. Sink an empty plant pot beside each plant to water in to. This allows the water to get to where it's needed without much evaporation or run off, and allows you to see how much water you're giving the plant (Figure 7.2).

4. Create small basins in the earth around plants, or raise the earth a little on the downhill side of plants in a sloping garden to help to slow run off of surface water when it rains (Figure 7.3).

5. Watering plants during the heat of the day can mean that up to 90% of the water simply evaporates. Water in the early morning or in the evening whenever possible.

6. Weeding will ensure that any water in the soil is available to your plants rather than the weeds, but if you don't like weeding, try applying mulch to suppress weed growth and conserve water at the same time.

Lawns

Do you really require your lawn to be green and lush, or can you put up with it going a bit brown for a few weeks in the summer? A happy medium may be to reduce its area, or replace it with some other surface (preferably an eco-friendly one like vegetable beds; not concrete or tarmac!). Increasing the organic content of the soil will help it stay green; organic matter retains water so top-dress the lawn every year with compost. Other tips:

1. Long grass will stay greener longer than short grass, and leaving cut grass on the lawn surface will allow it to act as mulch and help retain moisture.

2. Compacted lawns will tend to let surface water run off more quickly, so try and keep the top layer fairly open by spiking and then brushing compost into it.

3. If you're re-sowing your lawn, try and get seed that is drought-tolerant. Specialist mixes are available that contain high proportions of grasses such as creeping red fescue. New lawns should be planted in the autumn, and you should take care to make sure the soil contains lots of organic matter to help retain water; try and spread at least a couple of inches of compost over the entire area before you seed the lawn.

A garden sprinkler can use between 600 and 1400 litres an hour (compare this to the water uses illustrated in Chapter One). This is far more than grass needs, and much is lost to evaporation. Sprinklers also encourage surface roots, which will die off more quickly during dry periods. However, sprinklers remain the simplest and most commonly used technique for watering lawns. If you can't live without yours, get an electronic timer so you can water for an hour or so during the night, when evaporation is lower and demand on the mains network is reduced. The timer will allow you to programme when and for how long you water, and the sprinkler will be turned off automatically by a solenoid valve (see Figure 7.6).

Vegetable gardens

If you grow your own vegetables, you will need to keep them well watered, but you should try to group them according to their water needs. Root crops are the most

145

drought tolerant. Courgettes, squash etc may only need extra watering once the fruit starts to swell. Cauliflower, broccoli, salad onions, peas, potatoes, runner beans, tomatoes and leafy vegetables will need the most water, so should be put together somewhere where it's easy to water. Whilst this makes sense from the water efficiency perspective, it is not necessarily possible if you are carrying out traditional crop rotation. Also bear in mind that you can delay plant maturity and usefully spread the harvesting period by not watering some areas of the veg patch.

Trees

To help saplings through their first few years, you can sink a pipe a couple of feet into the ground next to them, so that you can easily apply a large volume of water, without allowing much run off or evaporation. Once a tree is established, it shouldn't need any extra watering.

Ponds

You can cut down the evaporation rate from your garden pond by allowing surface plants to cover as much as possible of the water surface. This will also help prevent it heating up (and since less oxygen is dissolved in warm water, the fish will prefer it cooler as well).

Alternative sources of water

The above measures will help make your garden as water efficient as possible, but you may still need to water your garden sometimes, and you may be interested in water other than that from your mains (or private) supply. If you are gardening commercially or on a large scale, you should refer to Chapter Two, which describes how to go about accessing large ground and surface water sources. This chapter is aimed more at basic techniques for the irrigation of domestic gardens.

It is inherently wasteful of energy and resources to use drinking quality water in the garden, but some care is needed if you are going to use lower quality water. This is particularly the case if you are watering food crops that you will then eat without first cooking them. Spray or sprinkler irrigation also poses a risk, since you may breathe in the water as airborne droplets, or swallow it. The following basic rules are offered as guidance for use of water in the domestic garden.

1) Neither rainwater nor grey water should be used in sprinkler or spray systems without prior disinfection.

2) Grey water should be applied below the soil surface wherever possible, and should never be stored for periods longer than 24 hours unless it has been treated.

3) You should avoid using grey water for irrigating crops that may be eaten raw, particularly if your grey water includes washing machine water from nappy washing.

Simple rainwater collection for garden watering

There are lots of environmental reasons why it makes sense to collect rainwater for use in the garden:

- water will be returned to the water cycle in the catchment area in which it fell, rather than being collected from a network of properties and discharged as storm water elsewhere
- dispersing rainwater to the ground close to where it falls means it won't get mixed with the dirtier water that runs off roads and paved areas
- decreased pressure on drainage infrastructure, particularly important in areas with a high proportion of combined sewers, which can discharge raw sewage to rivers in the event of heavy rainfall
- plants don't need drinking quality water; they will actually do better with rainwater than mains water
- reduces demand for drinking quality water in water scarce areas

Unfortunately, the time when your garden needs the most water is usually the time during which there is least rain. So we immediately have a mismatch between supply and demand, which as you will remember from Chapter Five, means that large storage volumes would be necessary to meet your entire demand. In reality, in most of southern England you will not be able to store enough winter rainfall to meet your summer gardening needs unless you store thousands of litres. Nevertheless, doing something is better than doing nothing! There are several levels of complexity possible, so you should decide fairly early on in the process how much effort you wish to go to.

1. Put water butts at the bottom of all rainwater downpipes. For many people, the limiting factor to how much rainwater they can store is more to do with the practicalities of siting water butts than actual storage volume. Slimline rainwater butts and stackable modular imitation-granite blocks are available. Agricultural and industrial tank suppliers have a wider variety of tank shapes and sizes than garden centres, and they are usually cheaper as well. Your local water company may also subsidise water butts. Consider re-routing gutters so that they can feed one large container, but if you do this you should ensure that the gutters will not be overloaded during heavy rain.

2. Make some simple modifications to your water butts (Figure 7.4). Put a tap

Figure 7.4 Rainwater butt for garden watering.

in the bottom of the water butt, and put the butt on a strong platform so you can get a watering can under the tap. An overflow pipe diverted to the existing rainwater drain will prevent the water butt overflowing from the top and causing dampness in any adjacent wall, and will make it easy to assess whether or not it is worth increasing your storage capacity. It's always worth putting a lid on your water butt, even if it's just a piece of wood. A lid improves safety, decreases evaporation, prevents contamination and helps stop potential pathogens breeding in the water by minimising exposure to light. This becomes more important as storage volumes (and therefore storage durations) increase.

3. If your garden topography allows it, run a hose-pipe from the tap at the bottom of a water butt to allow you to water the garden directly. The water pressure will be better using a tall water butt.

4. Rainwater filters and kits that divert water from your downpipe both remove solids like leaves and twigs from the rainwater and give you more flexibility in water butt placement. Removing solids is a good idea if you are planning an irrigation system, so that the holes in your irrigation pipe don't get blocked. If you're just filling watering cans, filters and diverter kits may not be worth the expense.

5. If you can calculate your garden water demand, then you can do the maths to work out how much water you should store. Refer to the calculations in Chapter Five, which will help you optimise your collection system.

6. In a domestic new build situation, or where you are having landscaping work done, you might consider large underground water storage tanks. You could use this stored water for toilet flushing in the winter and garden watering in the summer.

Domestic irrigation systems

The next level of complexity after simple garden hoses and watering cans is to install an irrigation system with a more permanent network of pipes. Such systems can be classified according to how the water is applied, and vary in their suitability for different water sources, as indicated in Table 7.2.

System	Water efficient	Water source irrigation technique is appropriate for:		
		Mains	Rain	Grey
Overhead mist	No	Yes	No	No
Sprinkler	No	Yes	No	No
Surface drip – commercial	Yes	Yes	No	No
Surface drip – DIY	Yes	Yes	Yes	No
Subsurface	Yes	Yes	Yes	Yes

Table 7.2 The type of irrigation system you use should depend upon what type of water you want to use in the system.

Overhead mist systems: consist of pipes held above ground level, dispersing water in mists or droplets. They are often mounted in polytunnel or greenhouse roofs. Sprinkler systems can be as simple as the common lawn sprinkler you attach to a hose, or more complex systems of a network of pipes and pop-up sprinklers buried underneath the lawn. You can even get 'walking' sprinklers that move themselves across the garden. Overhead systems and sprinklers are incredibly inefficient, as they allow massive amounts of evaporation. This is very temperature dependent, but can exceed 80% of the water applied under UK climatic conditions. Additionally, neither overhead nor sprinkler systems are suited to rain or grey water that hasn't been disinfected, since they result in airborne water droplets (and hence the risk of breathing in contaminated water), so they will not be considered further.

Figure 7.5 Trickle tape. This thin collapsible tape is an easy way of irrigating the garden. It is ideal for mains water irrigation systems, but because the holes in it are small it is not well suited to use with rainwater or grey water systems (courtesy of Access Irrigation).

On-ground (or surface) drip systems: these are basically pipes with holes in that allow water to drip onto the ground surface. Commercially available irrigation pipe for drip irrigation usually has a very small hole size; in soaker hose they are so small that the pipe appears to 'sweat'; trickle pipe is a thin collapsible polythene tube with slits in (Figure 7.5). Little water is lost by evaporation, and losses can be reduced further by covering the pipes with mulch. Neither is suited to rainwater or grey water, since the holes will block easily.

On-ground systems are very adaptable; the pipe can simply be moved from one bed to another.

Sub-surface irrigation systems: these consist of pipes with regularly spaced holes, laid in trenches that are then filled with mulch. They are a more fixed installation than on-ground drip systems, and are more work to install. However, they are more water efficient; no water is lost to evaporation, and water is applied directly to the roots of the plant rather than the ground surface. You can buy perforated pipe, or make your own by piercing holes in a piece of hose-pipe.

As indicated in Table 7.2, on-ground drip and sub-surface irrigation systems can be used with mains water. However, the latter are most water efficient and can be used with any water source, so are the best option for most gardens. The design process for a simple domestic irrigation system is described on page 155.

Automatic controllers: it's all too easy to forget to turn off a garden tap, and lose all the precious rainwater you have been collecting for months, or waste hundreds of litres of mains water. A timer that will turn off the tap after a fixed duration is therefore a good investment. The more complex timers will turn water on as well as off, and allow you to set a frequency and duration of watering (Figure

7.6), but you should bear in mind that any device that turns the water on could actually end up wasting more water than it saves, unless you also remember to deactivate it during times that the garden doesn't need watering. Devices that turn off after a set volume rather than a set time are also available, but owing to the way in which the switch works, they have a minimum operating pressure of around 0.5 bar, so are generally only suited to mains water irrigation systems.

Figure 7.6 Water timer. This device can be fixed to an outside tap to turn it on and off at times set by the user. This can reduce water wastage if you might otherwise forget to turn the tap off.

Reusing grey water in the garden

As discussed in Chapter Six, grey water is the waste water arising from showers, baths, basins, washing machines, sinks etc (i.e. all household waste water except that from the toilet). If you are following the principles of reduce, reuse, recycle, you will already have implemented water efficiency measures and your grey water production may be around 20-30 litres per person per day (as opposed to the UK average of around 100 litres), so you should think carefully about how much effort it is worth going to in order to reuse this fairly small volume of water. In addition to its low volume, grey water is often dirty and therefore difficult to deal with. You should certainly put effort into a rainwater system before concerning yourself with grey water. However, grey water is produced all year, so, unlike rainwater, it isn't going to run out exactly when the garden needs it most.

How 'dirty' is grey water?

Some of the things that make grey water 'dirty', such as soaps, shampoos, washing powder and cleaning agents may not be good for your soil. The components which are the main cause of concern are sodium, pH, phosphorous, and solids and fats.

Sodium

Sodium is an essential nutrient in the soil, but if levels are too high, the effect on your plants is similar to that of drought. The most significant source of sodium in your waste water stream (around 40%) is washing powders. The sodium is present as a filler or bulking agent, does not contribute to wash quality and yet can be the equivalent of pouring 90 grams of table salt down your drains every wash! Low sodium laundry detergents are available; these are typically concentrated liquid detergents or tablets as opposed to powdered products. Similarly, you should not use softened water in your garden, or the backwash water from a water softener, as this can also contain high levels of sodium.

pH

The pH of household cleaning chemicals, soaps and detergents tends to be very alkali, since grease and soiling is easier to remove at high pH. Plants that like acid soils may not thrive if irrigated with alkaline grey water. High pH in wash water can also reduce the life of clothes and cause skin irritation, as well as damage soil structure.

Phosphorus

Phosphorus is an essential plant nutrient, but too much of it is a bad thing, particularly in the aquatic environment where it results in eutrophication, a process in which algal blooms arise in response to the increase in nutrients and oxygen levels fall. So, whilst it won't cause your garden any problems, phosphorus has wider negative environmental impacts. The phosphorus load in sewage comes largely from detergents, and since compounds called zeolites can replace phosphorus in laundry detergents, it is unnecessary and in many countries its use is banned.

Solids and fats

Grey water contains varying amounts of solids, such as skin and hair, and kitchen sink grey water contains a lot of food solids and fats, all of which tend to block pipework.

Simple grey water reuse

For most people, grey water irrigation is best if it's kept simple. You may not want to reuse grey water from certain appliances, and it may be something you set up just for the summer. However, if you don't have a mains sewerage connection, a more elaborate system may be appropriate or even necessary. Grey water reuse can reduce the load on your septic tank, or with a more detailed design can be a treatment system in its own right; waste water treatment often includes a stage of filtering through gravel or sand in a tank, and plants can be grown in this filter medium, thereby combining gardening and sewage treatment. If you have a compost toilet, grey water reuse will mean you may have no need for a formal sewage treatment system at all. Interested readers are referred to *Lifting the Lid* and *Sewage Solutions* (see Further Reading).

At its simplest, grey water reuse consists of emptying the washing up bowl into the garden. There is much to recommend this minimalist approach; it has no construction costs, allows irrigation exactly where it is needed, and requires no maintenance. If you want to drain your bathwater out into the garden on an occasional basis, it's not really worth modifying your plumbing when you can simply drape a hose out of the bathroom window. You can set up a siphon without

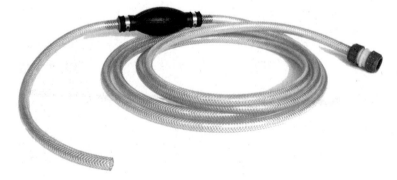

Figure 7.7 'DroughtBuster' siphon

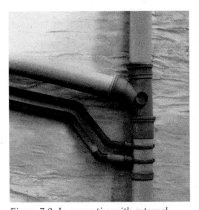

getting a mouthful of bathwater by using a 'DroughtBuster' (Figure 7.7) or a jiggle siphon (most often used for siphoning fuel, so look them up in automotive supply catalogues). Compare the average bathfull (80 litres) to standard sized water butts (200 litres) and you can see how this simple technique can help your garden through the summer when your rainwater butts are empty.

The next level of complexity involves some plumbing: separating your grey water waste pipes from your main soil stack. If you have a toilet upstairs, there will be a 110 millimetre wide pipe carrying the toilet water to the main sewer beneath ground floor level.

Figure 7.8 In properties with external sewage pipes (usually 110mm diameter), the pipes carring grey water from the baths and washbasins (usually 32 or 40mm diameter) may be fairly easy to disconnect and divert into a grey water irrigation system.

Smaller (32 or 40 millimetre) pipes carrying grey water from baths, showers, basins etc will be connected into this pipe at various points. These pipes can be located either inside the house in a duct, or outside the house attached to a wall. If the soil stack is on an external wall, then it will be easy to tell where the grey water joins and so reconnect it into a grey water system (Figure 7.8). You can either make this reconnection permanent, or use a 'Water Two' device (Figure 7.9) that incorporates a valve operated by pull cords to divert the grey water. If you have an internal

Figure 7.9 'Water Two' device incorporating a manually operated valve to divert grey water.

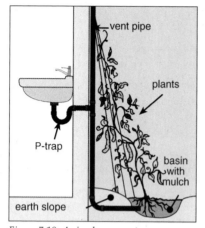

Figure 7.10 A simple grey water reuse system for washbasins.

soil stack, then diverting the grey water might require drilling through external walls, which could be quite a lot of bother in proportion to the small volumes of grey water produced by a water efficient household.

A simple system for washbasins

Washbasin waste water is produced in small quantities and can be reused using the system illustrated in Figure 7.10. The outlet pipe can travel any distance from the house, provided it has at least a 2% fall. Depending on the layout of your garden, it may be possible to have several of these systems, one for each washbasin. This system isn't so good for kitchen sink water, which contains a lot of solids and grease and which will tend to block any pipe-work.

Whole house grey water systems

If you wish to collect grey water from the larger volume sources (such as showers, baths, washing machines), you will need to incorporate a simple filter and surge tank. The system shown in Figure 7.11 will remove solids (and so minimise the risk of blocking any irrigation systems), allow the water to cool and temporarily store some of the water so that household appliances drain quickly. A tank of around 40 litres will be sufficient, unless you have a particularly high grey water output. Straw makes a good filter medium and can be removed and composted every few months.

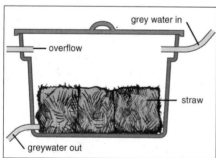

Figure 7.11 Grey water surge tank to filter grey water for garden use.

Having filtered the grey water and evened out the flow a bit, you need to consider how and where to distribute it in the garden. As stated earlier, grey water should be used within 24 hours, because it will quickly stagnate and become smelly, so a system that automatically discharges into the garden is preferable to one in which handling is required. It is also better if applied below ground level in a sub-surface system; on no account should untreated grey water be used in spray form. Design of a sub-surface irrigation system for grey water is described below.

Irrigation system design

Having decided on a source of water and how best to move it to where it's needed, you need next to consider the irrigation system itself.

1) Draw a plan of the garden and decide where you wish to water.

2) Rainwater only: where are you going to store the water, and how are you going to get it from the tank to the area to be irrigated? The higher up you can store the water relative to where you want to irrigate, the better the system will perform. The relationship between water pressure and flow is complicated, but as a rough guide you will need to have at least a metre of height difference for every 5 metres of 16 millimetre diameter dripper pipe.

3) Will it be an on-ground system or a sub-surface system? On-ground systems are simpler to install and make it easier to move the pipe around, but sub-surface systems are more water efficient and allow you to incorporate grey water as well as rainwater.

4) Work out the most efficient way of laying out the pipes. The water pressure will drop along the length of the pipe, so less water will be delivered through the holes further away from the supply end. You should therefore try and have a main pipe with branches, rather than a single length of pipe (Figure 7.12). This design will also allow you to have taps on the branches, so you can concentrate the watering where it is most needed. Limiting the distribution to individual branches at any

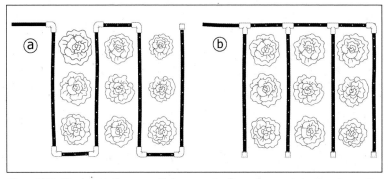

Figure 7.12 Pipe layout. Pressure will drop as water flows along an irrigation pipe, so it makes sense to have a main pipe with several branches coming off it (as in b), rather than a single continuous pipe (a).

Figure 7.13 On-ground (left) and sub-surface (right) irrigation systems.

one time will also help if you have low water pressure. The end of each pipe should have a stopper in it to improve the distribution of water. You can remove this stopper periodically to flush water through the pipe-work to clear any blockages. Slopes will also affect the distribution (commercially available pipes have pressure compensation nozzles to reduce this effect, but the nozzles are liable to block if used with rain or grey water).

5) Work out how many connecting pieces you need and of what type (T pieces, elbows, blank ends etc), and how much pipe you need. You may also need some specific plumbing parts like check valves or 'Type DB valves' to minimise the risk of any rain or grey water entering the mains water system (see page 157). You can make your own grey water irrigation pipe out of hose-pipe by piercing holes in it, or out of plastic waste pipe (32 or 40 millimetre). Hose-pipe has the advantage of flexibility, so you won't need so many joints, but is more difficult to lay flat. There are no hard and fast rules on how big the holes should be and how far apart; it is very dependent on the length and diameter of the pipe and the incoming pressure. At CAT we have used 32 millimetre waste pipe with 8 millimetre holes every 30 centimetres, and half inch hose-pipe with 3 millimetre holes every 20 centimetres

Figure 7.14 Protruding distribution pipe for a grey water system.

also works well.

6) Install the pipe-work. Sub-surface systems are laid in 10-20 centimetres deep trenches alongside each row of vegetables in the garden (Figure 7.13), and the trench backfilled with woodchip to act as a mulch. Using mulch-filled trenches rather than laying the pipes directly into the earth reduces the risk of root penetration and soil blocking, and allows pipes to be easily removed and cleaned should they become blocked.

The woodchip composts and is replaced every couple of years. If you are planning a number of lines of pipe, you can make the system more adaptable by allowing the end of the pipe to emerge above ground level, and simply put the end of the hose from the surge tank into whichever pipe run you wish to use at the time (Figure 7.14).

Pumping water for garden use

Electrically powered pumps suitable for moving hundreds of litres of water per day are dealt with in Chapter Three. If you want to pump small volumes of rain or grey water for the garden you can often do it by hand. For example, you may want to pump rainwater from a water butt at the front of the house to a more convenient one at the back. Hand operated bilge pumps (Figure 3.13, page 64) connected to garden hoses are well suited to this purpose, and are available from around £40. Most will pump to around 3 metres height, and you can pump 40 litres a minute without too much physical effort. If you just need to set up a siphon, the DroughtBuster (Figure 7.7) is your best bet.

A word of warning; you may be tempted to connect the outlet hose of your washing machine to a grey water system and use the power of the washing machine pump to move the water up a gradient into an irrigation system. These pumps are not designed to take this type of load, so if you want to use the grey water arising from your washing machine, you should discharge it via a conventional standpipe and trap. This in turn may be connected to your grey water system if the gradient is suitable.

Regulations that apply to water in the garden

It is important to minimise the risks of contaminated water from the garden being drawn back into the mains water supply. This can easily happen in the garden; for example if you are filling a water butt with a garden hose and the pressure in the mains water supply drops, water from the water butt will be drawn back into the supply pipes. Measures required under the Water Supply (Water Fittings) Regulations are designed to prevent this type of event.

Outside taps

A 'double check valve' is required either in the supply pipe (in new installations), or within the tap itself (where an outside tap is being replaced). Suitable taps will say 'HUK1, HA, EC or ED' on the packet. These basically prevent water flowing the wrong way. Similarly, you shouldn't refill water butts from taps indoors using a hose, unless they have some kind of backflow protection.

What to do during a hosepipe ban or drought order

Hosepipe bans and drought orders are becoming increasingly common, particularly in the South East. A hosepipe ban makes it illegal to use a hosepipe or sprinkler attached to a mains water supply tap for watering private gardens, or to use a hosepipe for washing cars. They do not restrict your use of rainwater stored in water butts, or grey water, or prevent you from using a hosepipe that isn't attached to a mains water supply tap. Neither do they prevent you from watering with a watering can. So, you can use hosepipes to move stored water around (although you might want to point out to any anxious neighbours exactly what you are doing so that they don't try and report you) and you can continue to use hosepipes/sprinklers from private water supplies such as boreholes, although it would be pretty irresponsible and un-environmental to do so.

Drought orders are more extreme than hosepipe bans, and the limitations on water use will vary, but will be well publicised by the water company concerned. For the most part, they don't affect householders and are aimed at large scale users. However, they will often include limitations on the filling of private swimming pools and ornamental ponds.

If you've planned ahead for an impending hosepipe ban, you will doubtless have some rainwater saved in butts. However, most of us won't be able to store as much as the garden needs, so you may well decide to start using your grey water as discussed earlier in this chapter. Keep it simple to start with; just emptying your washing up bowl into a flower bed (a handy 10 litres), or siphoning water out of the bath (around 80 litres), doesn't require you do do any additional plumbing and will help your plants through the worst of any dry spell. Sinking plant pots into the ground, or creating little mounds of soil around plants as discussed on page 144 will enable the water to soak into the soil rather than simply running off the surface. The next thing to do is to start designing an irrigation system and installing rainwater butts ready for the following summer.

Irrigation systems connected to a tap require the tap to have a double check valve as described above, but also a 'Type DB device', which is a more failsafe backflow prevention measure. NB: More stringent precautions are necessary in commercial installations; you should consult the WRAS Water Regulations Guide, or an irrigation engineer if you are unsure what is required.

Summary

There are a number of ways of making your garden water efficient, and additional water should be collected from roofs in preference to the use of mains water. DIY irrigation systems can be used to optimise the use of this water, and can be adapted to include use of grey water from baths, showers and washbasins. If you only really need to water your garden for a couple of months in the summer and/or during a hosepipe ban, simple grey water recycling techniques should be sufficient.

Further Reading

- *The Dry Garden*, Beth Chatto, Orion Books, ISBN 0-460-02222-9, 1998. Advice on choice of species and garden design in dry areas
- *Garden watering systems*, Susan Lang, Sunset Books, 1999. Irrigation system design and installation
- *Lifting the Lid: An ecological approach to toilet systems*, P. Harper & L. Halestrap, CAT Publications, 1999. Available from CAT Mail Order.
- *Sewage Solutions: Answering the call of nature*, Grant, Moodie & Weedon, CAT Publications, 2002. Available from CAT Mail Order.
- *Watering systems for lawn and garden, a do-it-yourself guide*, R Woodson, Storey Books, 1996. Irrigation system design and installation
- www.rhs.org.uk/advice The Royal Horticultural Society has lots of useful information on water efficient gardening and drought tolerant plants. The RHS garden at Hyde Hall has an area dedicated to drought tolerant plants.
- www.oasisdesign.net/greywater has a wealth of information on grey water use in the garden (an American site)

Conclusions

The largest single measure that can be undertaken to preserve the life and health of a population is to install safe water supply and sewage treatment systems. In the UK we have a safe, robust system that provides millions of litres of clean water a day and treats our sewage. Any system of this scale clearly has some environmental impact, and as responsible citizens we should do all we can to modify our behaviour to reduce this impact. However, our impact due to water and sewage is one of the lesser of many evils, comprising an average of just 2% of our ecological footprint. Consequently, for most people whose primary motivation is the environment, it is not appropriate to prioritise water above energy, transport or food miles.

Nevertheless, there is a lot we can do. Since using less of a resource is easier than finding more, water efficiency should be the first resort and there are many behavioural changes that will help you reduce your water consumption. If you are upgrading water-using appliances, you should bear in mind their water efficiency. Collecting rainwater in butts for use in the garden is also a good strategy, as it goes some way to alleviating problems with surface water drainage in the winter, and also problems with droughts and hosepipe bans in the summer. In new build properties, or those with larger roof areas it may be worth looking at more elaborate rainwater harvesting systems, possibly including reuse for toilet flushing in addition to use in the garden. Hosepipe bans are a regular feature of the English summer, and planning ahead will enable you to go a long way towards preventing your plants from suffering too much. These measures may well include limited recycling of 'grey' water from showers and baths.

The merits of installing a private or supplementary water supply will vary widely between households and we would urge you to be mindful of the bigger picture before trying to become independent from mains water. As discussed in Chapter One, this should include considerations of 'virtual water'. However, in some locations mains water isn't an option. Finding your alternative water source will be your first challenge. In some instances there will be a variety of sources available and the challenge will be the decision making process as to which one to use. In other instances, the chance of finding a water source may seem remote. An experienced borehole drilling company may be able to help you at this point. Once you have found your water, you need to make it safe to drink. The way in which you clean water depends on what is in it that you are trying to remove; it is best to start from water test results rather than a preconceived idea about what type of filter you would like.

There are many reasons why you may wish to install a private water supply system other than considerations of environmental impact. However, it is not a decision to be taken lightly - given the risks to your health from a poorly functioning system. No private water supply system will be 'fit and forget'; if you are not the type of person who is likely to get round to checking the system and doing maintenance when required, make sure you have an arrangement in place to make sure that it does get done. Having read this book, you will be in a better position to make an informed decision on suitable water sources, treatment methods and upkeep whilst maintaining an environmental perspective.

Glossary

Alkaline – of high pH (the opposite of acid)

Anion – a negatively charge ion (e.g. Cl^-)

Aquifer – an underground area that can act as a water store (e.g. sand, limestone)

Artesian well – a well whose water level is higher than that of the surrounding ground water owing to a high hydraulic pressure within the aquifer

Bladder storage tank – a type of pressure storage tank containing a water filled bladder

Cation – a positively charged ion (e.g. Na^+)

Coagulation – the clumping together of fine particles into larger particles (which are then easier to separate from the water)

Coliform – a type of bacteria that is often used as an indicator organism; a sign that dangerous micro-organisms may be present

Diaphragm storage tank – a type of pressure storage tank in which a rubber membrane separates the gas filled compartment from the water filled compartment

Ecological Footprint Analysis – a technique used to measure the environmental impact of a product or activity, based on the amount of land area required to support that product or activity

Flocculation – a method to remove the suspended solids from water by causing them to stick together and therefore sink faster

Flow restrictor – a valve designed to limit the amount of water that can flow through a pipe

Grey water – waste water from household sources with the exception of toilet water, i.e. showers, washbasins, sinks, washing machine, dishwasher

KDF – a copper-zinc alloy used to remove a range of contaminants from water

Life Cycle Analysis – a process by which the environmental impact of products can be compared, by adding up their environmental impacts during manufacture, use and disposal

Pathogen – a micro-organism that may be dangerous to health

Perched aquifer – an isolated area of groundwater with a small recharge zone, rarely suitable for a private water supply owing to its likelihood to run dry

pH – a logarithmic scale of acidity. High pH is alkali, low pH is acid

Pressure storage tank – a water storage tank that is designed to minimise pump cycling by allowing water to fill and empty over a range of pressures, similar to a pressure expansion vessel in a sealed heating system

Protozoa – a type of micro-organism, often living parasitically (e.g. Giardia, Cryptosporidium)

Rapid gravity sand filter – a water purification technique which removes solids from large volumes of water relatively quickly, with the disadvantage that energy is required for frequent backwashing to remove the solids from the filter

Residual disinfection – the continued presence in water of a compound that can kill pathogens (e.g. chlorine)

Reverse osmosis – a water purification technique in which water is passed under pressure through a membrane with very small holes in it

Slow sand filter – a water purification technique in which water flows by gravity through sand, which acts as a filter to remove solid particles and micro-organisms

Specific heat capacity – the amount of heat needed to raise the temperature of water

Submersible pump – a pump that operates under water by pushing the water through the delivery pipe

Suction pump – a pump that operates above water by sucking water into itself

Transpiration – process by which plants lose water to the air through the pores in their leaves

Upflow sand filter – a water purification technique in which water is pumped upwards through sand to remove solids

Virus – a disease causing micro-organism without a nucleus

μ – Greek letter known as 'mu'; symbol meaning micro (as in μg or microgram)

Acronyms

BSRIA – Building Services Research and Information Association
DBP – disinfection byproduct
DWI – Drinking Water Inspectorate
EA – Environment Agency
EH – Environmental Health
EHS – Environment and Heritage Service (the EA equivalent in Northern Ireland)
MDPE – medium density polyethylene (a type of plastic pipe)
NAMAS – National Accreditation of Measurement and Sampling
NSF – National Science Foundation (in America)
RO – Reverse Osmosis
RWH – rain water harvesting
SEPA – Scottish Environmental Protection Agency
UKAS – United Kingdom Accreditation Service
WHO – World Health Organisation
WRAS – Water Regulations Advisory Service
WRc – Water Research Centre

Appendix

Changes in the Private Water Supply Regulations

Since writing the first edition of this book (in 2005), regulations governing private water supplies have changed. The drivers for this include the European Drinking Water Directive, modified WHO guidelines and the report of a 2001 Government Task Force on *E. coli*. In Scotland, the Private Water Supply (Scotland) Regulations 2006 have replaced the 1992 Regulations. The regulations for England and Wales, and Northern Ireland, will also change, probably in 2007/2008, and are likely to be very similar to the 2006 Scottish Regulations. If you are in any doubt about which regulations are in force, consult your Local Authority Environmental Health Department. The following describes the Private Water Supplies (Scotland) Regulations 2006.

Categorising water supplies

Private water supplies are categorised according to the number of people served, and whether the water is for public or commercial use. These categories are shown in Table A1.

Type A	Type B
Supplies on average at least 10m^3 of water per day *OR* serves 50 or more persons *OR* is supplied or used as part of a commercial or public activity	Any supply that does not fall within the definition of a Type A supply. i.e. supplies less than 10 m^3/day, where the water is solely for private use

Table A1 Types of water supply under Scottish Regulations.

Within Type A supplies, there are three 'levels' of supply according to the volume of water supplied (Table A4). These influence the frequency at which the Local Authority (LA) will test the source, but have no effect on the parameters tested for.

Type B supplies

If you are responsible for a Type B supply, you are required to tell the LA about it, and they are required to keep a record of it, along with basic details. The LA has the right (but not the duty) to monitor Type B supplies on an ongoing basis. The extent to which this is done will therefore vary from region to region.

Changes in parameters tested for

The water quality parameters tested for in the 2006 regulations are broadly similar to the basic parameters in the 1991/92 regulations and are given in Table A2. In addition, the LA may test for other parameters the presence of which is suspected and which may cause harm to the health of those using the supply.

Parameter	Unit of measurement	Concentration or value (maximum unless otherwise stated)	Old requirement
Hydrogen ion	pH value	6.5 minimum, 9.5 maximum	5.5 minimum, 9.5 maximum
Conductivity	µS/cm	2500 at 20°C	1500 at 20°C
Total coliforms	number/100ml	0	0
Faecal coliforms	number/100ml	N/A	0
Enterococci	number/100ml	0	N/A
E. coli	number/100ml	0	N/A
Lead	µg/l	25: until end 2013; 10: from 2014 onwards	50
Nitrate	mg/l	50	50
Odour	Qualitative		
Taste	Qualitative		
Turbidity	Formazin turbidity units	4	4

Table A2 Required standards for parameters that will be tested under the Private Water Supply Regulations. The LA has the discretion not to test for nitrate if they believe the levels in the locality are below 25mg/l.

The rationale for some of these changes is indicated below:

Enterococci

The inclusion of these bacteria reflects the fact that they persist longer in the natural environment than many coliform bacteria and are more resistant to treatment. It is therefore a more conservative indicator of microbiological contamination than total coliforms. Removing micro-organisms from water is discussed in Chapter Four, page 86.

E. coli

The inclusion of *E. coli* as a parameter reflects the fact that the infectious dose (the number of bacteria necessary to produce an infection) is extremely low compared to many other bacteria.

Lead

The gradual decrease in allowable lead concentrations reflects concerns from epidemiological studies on the long-term effects of lead. Removing lead from water is discussed in Chapter Four, page 94.

pH

The narrowing of permissible pH values reflects the fact that acidic water is corrosive and can dissolve metals, both in the natural environment and in your domestic plumbing system (which might not get picked up in a water test performed on your kitchen tap since it may be occurring in other parts of the plumbing system). pH modification is discussed in Chapter Four, page 93.

Risk assessment

By far the biggest change in the regulations concerns the use of risk assessment. A water sample is no indication of past or future water quality since contamination of a water supply is often sporadic, e.g. only occurring after heavy rain, or dependent on the presence of infected farm animals in neighbouring fields. A study in Scotland found that the majority of private water supplies contained microbiological contamination at least once a year, but many of these supplies were clear most of the time. Since many supplies are only sampled annually (if at all), significant numbers of people are subject to risk from their water supplies and may have a false sense of security if they regard a test result as a permanent 'all clear'. Risk assessment is therefore useful because whilst the presence of microbiological contaminants is intermittent, it can be predicted from the level at which the water source and treatment system is operated and maintained.

You can carry out a simple risk assessment yourself, with forms downloadable from the website (www.privatewatersupplies.gov.uk), or available from your LA, along with guidance notes on how to complete them. The assessment consists of simple yes/no questions relating to the presence of source protection (the type of measures discussed in Chapter Two). It explains in simple terms why each type of protection is necessary. This risk assessment is sent back to the LA, who may then choose to visit and complete a full assessment.

Again, the full assessment has a set of standard forms which differ slightly depending on whether the supply is a well, borehole, spring or surface water supply. The forms have 4 main sections:

• **Section A** – general questions on supply category, number of people served, type of purposes water is used for, contact details of person in charge of supply etc.

Section	Types of question
D (i) Site survey	Presence of likely pollution sources such as wildlife, livestock, waste disposal sites, sewerage systems, farm wastes, industrial activity, pesticide use etc.
D (ii) Supply survey	• State of repair of source. Particular emphasis on source protection such as fencing and manhole covers. • Condition of distribution system such as header tanks and pipework. • Suitability of maintenance regimes and treatment techniques.
D (iii) Soil leaching risk survey	Only present on risk assessment forms for springs or wells. Certain soil types and underlying geology can protect groundwater sources from surface contamination by their filtering action, whereas other soil types allow contaminants directly into groundwater.
D (iv) Overall risk assessment	The final recorded risk category

Table A3 Types of questions on official risk assessment forms.

- **Section B** – diagram and description of supply. Daily volume abstracted, details of water treatment processes etc.
- **Section C** – previous sample results, enforcement notices, previous risk assessments etc.
- **Section D** – the detailed risk assessment (See Table A3).

The risk assessment takes into account both the severity of the hazard and the likelihood of it actually occurring in order to come up with an overall risk score.

In practice, the full risk assessment forms are only really suited for use by professionals. Your local Environmental Health Officer will carry out the assessment; they will then advise on measures that you can take to reduce your risk. If you want to make improvements to your private water supply without involving the 'authorities', then the form will help you to determine what type of measures you should take. In most instances these will be related to source protection (discussed in Chapter Two) or treatment techniques (discussed in Chapter Four).

Type A supplies

These are supplies where either the supply is more than 10m³/day, or serves more than 50 people, or is used for commercial or public activity (table A1). This

Level	Maximum volume supplied (m^3/day)
1	≤ 100
2	> 100 ≤ 1000
3	> 1000

Table A4   Levels of Type A water supplies.

definition therefore includes accommodation providers within the tourist industry. These supplies are classified into 'levels' (Table A4), which in turn determine the frequency at which the authorities will analyse water samples from them. Level 1 supplies must be sampled annually, level 2 supplies quarterly, and level 3 supplies at a higher frequency (derived from the total volume of water supplied).

Types of monitoring

The regulations on Type A supplies also differentiate between check monitoring (for parameters which must be complied with at all times, as they have the potential to cause immediate harm to health) and audit monitoring (which applies to parameters that will affect health if the standard is continuously breached). The parameters requiring check monitoring are listed in the regulations (available on the web, see Further Reading for details) and are broadly similar to the parameters listed in Table A2. Those subject to audit monitoring are broadly similar to the additional parameters detailed in Table 4.5 on page 78, Chapter Four. The importance of this distinction between check monitoring and audit monitoring is in the issuing of derogations, discussed below.

Interim arrangements and derogations

Inevitably, some Type A private water supplies will not fully comply with regulations, particularly when new regulations have recently come into force. Provision is made to allow temporary departures from the required standards, but only in the case of those parameters that are not considered immediately dangerous to health; the ones that are subject to audit monitoring as opposed to check monitoring (discussed above). These derogations will be for fixed periods of time, during which improvements must be made to the water supply to bring it up to standard. The Environmental Health Officer will advise on this process.

Associated regulations
Additional sets of regulations cover the availability of grants for improvement work to private water supplies, and there is a requirement for any owner of a Type A supply to put up a prominent notice on the premises informing the public that a private water supply is in use.

Conclusions
The new Private Water Supply Regulations are a considerable improvement on their predecessors, owing to the emphasis on risk assessment and source protection, as opposed to simply relying on infrequent water testing. The parameters tested for also more closely reflect the risks to health from private water supplies. They will also assist both owners of private water supplies and Environmental Health Officers to have constructive dialogue on improvements to supplies and safeguarding health.

Further Reading
- www.privatewatersupplies.gov.uk   Guidance on the Private Water Supplies (Scotland) Regulations 2006. Will incorporate guidance on changes in regulations in Northern Ireland, England and Wales as they occur.
- *The Private Water Supplies (Scotland) Regulations 2006*, ISBN 0110702840. Also available on the HMSO website.

Index